Preface

Graphene, the leading example of a two-dimensional crystalline material, is exceptionally conductive of electricity and heat even at 1 atom thickness, taken as 0.34 nm. This short book, in the series SpringerBriefs in Materials Science, is intended to survey emerging applications of the material that has received great attention since its isolation in 2004 by Andre Geim and Konstantin Novoselov. These brilliant experimental physicists shared the Nobel Prize in Physics in 2010 for their discoveries. Following introductory information on the material and conduction processes that occur in it, we discuss methods of making graphene in varying degrees of perfection and cost. Main application areas are discussed, with emphasis on areas including solar cells and computer logic devices, as well as niche applications within the semiconductor technology, that potentially will have large impact. I am grateful to Christopher T. Coughlin, Physics Editor of Springer Science and Business Media, in New York, for suggesting a short book on *Applications of Graphene*. I thank Ms. Ho Ying Fan at Springer for her rapid processing of my manuscript. I have benefitted from the support of Prof. Lorcan Folan in the Applied Physics Department at NYU Poly. I thank Ms. DeShane Lyew, of the Applied Physics Office, and Mr. Malhar Desai, for assistance in preparing the manuscript. My wife Carol has been a source of continuing help and encouragement in these projects.

Brooklyn, USA, September 2013   E. L. Wolf

Contents

Contents

Chapter 1
Physical and Electrical Properties of Graphene

1.1 Introduction

Graphene is the single layer of the graphite crystal, pure covalently bonded carbon in a honeycomb lattice, one atom thick. The single layers can be detached from graphite, and grown by conventional chemical vapor deposition methods. The single layers are extremely strong in tension and bend very easily, but do not tear. They are remarkably good conductors of electricity and heat. To a first approximation graphene is a semimetal, and certainly has no band gap. The Fermi level, lying at the intersection of conduction and valence bands in pure material, can be shifted to make it N- or P-Type, by chemical additions, or, most easily, by an electric field. The electron energy bands near the neutral point, termed the Dirac point, are linear, rather than parabolic, in their dependence on wave vector k. The new electron properties arise in a straightforward way from the symmetry of the atomic positions, resulting in cone-shaped, rather than parabolic, regions in the electron energy surfaces. All the new effects are well described by the Schrodinger-equation-based methods of condensed matter physics, i.e., the Dirac-equation-like electron behavior is obtained directly from appropriate simplification of the Schrodinger theory of solids. The linear bands mean that there is a single velocity, v_F, approximately 10^6 m/s, for all carriers within a few tenths of an electron-volt (eV) of the neutral point energy. The lattice symmetry forbids direct backscattering of electrons, enhancing the electrical mobility and conductivity. The subtlety of the band structure brings in analogies to phenomena known in high energy physics. The carriers formally behave as spinors, but the impact, beyond the improved mobility, of these interesting and confirmed effects on applications seems modest. In application, the main novelty of graphene is that it is fully functional and structurally continuous, displaying exceedingly high electrical conductivity, at one atom thickness. This fact is hard to believe but is well confirmed. Electron device applications can exploit the tunability of the Fermi level and thus the work function, by electric field. Applications can also benefit from the basic symmetry of the electron properties above and below the neutral point. This symmetry is a positive aspect in considering graphene-based devices as a

E. L. Wolf, *Applications of Graphene*, SpringerBriefs in Materials, DOI: 10.1007/978-3-319-03946-6_1, © Edward L. Wolf 2014

replacement of the CMOS (Complementary Metal Oxide Semiconductor) field-effect transistor logic. Graphene is a fully covalent conductor based on extremely strong sp^2 trigonal bonds between carbon atoms. The graphene structure fully engages the four n = 2 valence electrons of carbon: three of these covalently bond the triangular lattice, the fourth electron, in a $2p_z$ state, gives the electronic conduction. The full covalent bonding leaves graphene as chemically inert, extremely strong, and refractory, while the single free p_z electron gives exceptional electrical conductivity. The parent graphite crystal sublimates at 3,900 K, suggesting that intact graphene layers detach at 3,900 K. Disintegration of the individual layers has been variously estimated to occur at temperatures 4,900 and 5,800 K. The profound strength and refractory nature of this structure makes it difficult to sinter small platelets of graphene (that are easily available) to achieve larger single crystal electrodes. From the point of view of application, a practical difficulty is that there is no analog of the nearly defect-free 300 mm silicon wafer, the basis for contemporary high-level chip (and solar cell) manufacture. The closest graphene analog of the large wafer of silicon is a large sheet of graphene grown at 1,000 °C by catalytic thermal decomposition (CVD) of methane on copper foil. This method has allowed continuous sheets of 30 in. diagonal measure, containing large 2D grains. If four such sheets are chemically doped and stacked, a nearly transparent continuous flexible electrode of sheet resistance 30 Ω/square has been achieved (Bae et al. 2010).

The mechanical properties of graphene are well-predicted by classical bending-beam formulas, (Bunch et al. 2007) taking beam thickness as t = 0.34 nm, the layer spacing in graphite, and a Young's modulus Y of 1 TPa. For example, a classical spring constant K, representing the force F = Kx needed to deflect a square of side L, clamped on one edge, by a distance x, is given as $K \sim Yt^3/L^2$ (Ekinci and Roukes 2005). For L = 1 mm, this gives K = 39 N/m, but for L = 1 micrometer, $K = 39 \times 10^{-6}$ N/m. Thus, a graphene square of molecular size is stiff, but for size L larger than a micrometer, it has vanishing stiffness, and will closely adhere, by van der Waals force, to any substrate, and will thus acquire the roughness of that substrate. An interesting experiment has shown that a monolayer graphene membrane is impermeable to helium atoms (Bunch et al. 2008). The membrane held an overpressure of helium gas for hours, making clear that it was indeed continuous, defect-free, and would not allow the smallest atom to pass through. This property is easily confirmed by elementary tunneling theory, for example see Wolf (2013).

Another property of graphene is its vast surface area, 2,600 m^2/g. The surface area per gram is maximal because graphene is one atom thick, the only other possibilities would be Li, Be, and B (mass numbers 7, 9, and 11) versus mass number 12 for C. But none of these forms continuous sheets, although the graphene analog BN does form the same hexagonal 2D structure (it is an insulator). Another way of stating this is that graphene is all surface, with no bulk. So, while it is generally true that graphene is inert, when one looks more closely, physisorbed atoms and molecules are common and do substantially reduce the electrical conductivity of graphene. The conductivity of graphene can be increased, by factors exceeding two, by gentle heating to ~ 400 °C in modest vacuum (Bolotin et al. 2008) (Molecules can also increase the conductivity by adding carriers to graphene, but the mobility is always reduced.). The surface sensitivity of graphene, also present in carbon nanotubes, is the basis for its use as a sensor, even, of single molecules (Schedin et al. 2007). In practical terms, graphene is hard to deal with, it has no rigidity on scales larger than micrometers and it is nearly invisible. The optical transmission of one monolayer is 97.7 % (Nair et al. 2008), and a basic underlying reason for the Nobel Prizes for A. Geim and K. Novoselov in 2010 was (Geim 2011); (Novoselov 2011) that they found a practical optical approach to clearly identify a single monolayer of graphene, placed on an oxidized Si surface. Graphene is the leading example of a 2D material (BN being a second) that can be fabricated in three-dimensional space. A mathematical literature establishes that a 2D layer, even if it is completely constrained to a plane, has some limitation on its lattice vibrational properties. The fortunate situation is, apart from admitted difficulty in fabricating large samples of graphene and BN, that the mathematically predicted limitations on in-plane vibrational properties are so small as to have escaped detection. The practical problems of fabricating and handling graphene and similar 2D materials, in suitable size and quality, remain formidable.

1.2 Basic Lattice and Electronic Structure

The lattice is shown in Fig. 1.1, where A and B indicate identical carbon atoms but located on the two interpenetrating triangular lattices that make up the honeycomb lattice. In this figure, $\mathbf{a_1}$ and $\mathbf{a_2}$ are the basis vectors that generate the lattice, while $\mathbf{b_1}$, $\mathbf{b_2}$ and $\mathbf{b_3}$ are the nearest neighbor translations. In more detail, we have basis vectors $a_1 = (\sqrt{3}/2, -1/2)a$, $a_2 = (0, 1)a$, and the sublattices are connected by $b_1 = (1/2\sqrt{3}, 1/2)a$, $b_2 = (1/2\sqrt{3}, -1/2)a$, $b_3 = (-1/\sqrt{3}, 0)a$, in terms of the nearest neighbor distance, $a = 142$ pm, where 1 pm $= 10^{-12}$ m. The bonding in this structure is of the planar covalent sp^2 type based on the $n = 2$ electrons of the carbon atom. An important aspect of this structure is that nearest neighbor atoms are on different sublattices, denoted A and B. The corresponding Brillouin zone is depicted in Fig. 1.2.

Viewing Fig. 1.2, with nearest-neighbor distance $a = 142$ pm, the lattice constant is $3^{1/2}a$, and the zone boundary M (half the reciprocal lattice vectors $\mathbf{b_1}$, $\mathbf{b_2}$), is $2\pi/3a$. The coordinates of the corner point K are $(2\pi/3a, \pi/3\sqrt{3}a)$ so that the distance from the origin to point K is $4\pi/(3\sqrt{3}a)$. Since the conduction and valence bands touch precisely at K, we have $k_F = |K|$ and the Fermi wavelength $\lambda_F = 2\pi/k_F = 3\sqrt{3}a/2 = 369$ pm.

The electron bands that arise in this lattice were first calculated by Wallace in 1947, who realized that the bands of the single plane that he calculated were a good approximation to the bands of graphite, since the planes are so weakly coupled and so widely spaced, by 0.34 nm.

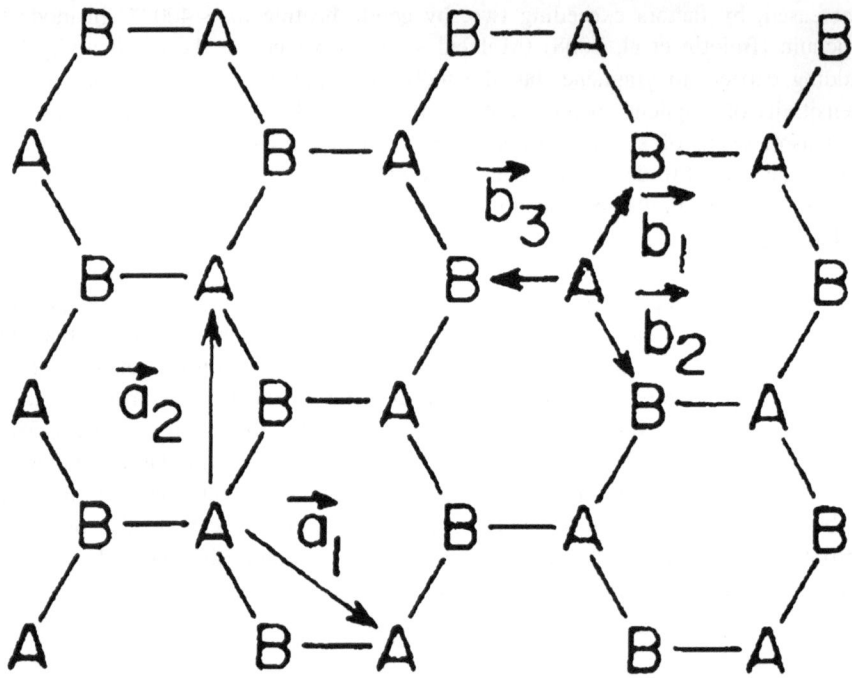

Fig. 1.1 *Honeycomb lattice*, resulting from A and B interpenetrating *triangular lattices*. The nearest neighbor distance is a = 142 pm, and the lattice constant is $3^{1/2}$ a = 246 pm. (Reprinted figure with permission from Semenoff et al., Fig. 1. Copyright (1984) by the American Physical Society)

Fig. 1.2 Hexagonal Brillouin zone of *honeycomb lattice*, resulting from A and B interpenetrating *triangular lattices*. Reciprocal lattice vectors are $\mathbf{b_1}$, $\mathbf{b_2}$ and the essential Dirac points are K, K'. (Reprinted figure with permission from Castro Neto et al., Fig. 2. Copyright (2009) by the American Physical Society)

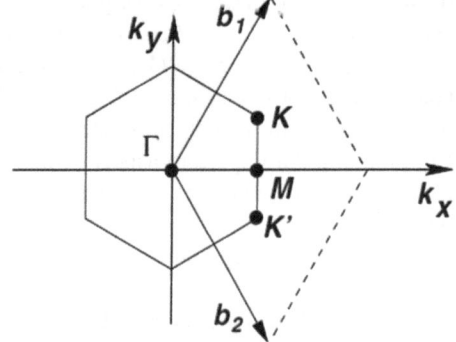

A modern representation of that bandstructure is shown in Fig. 1.3. The upper and lower bands are derived from the $2p_z$ orbitals of carbon. The conical crossings at K and K' (Wallace 1947) are the result of the two-sublattice symmetry.

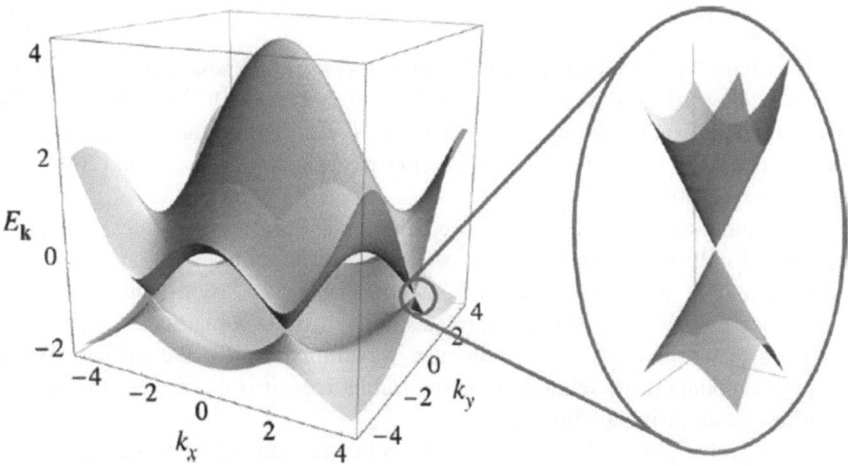

Fig. 1.3 Energy bands of graphene, essentially as given by Wallace (1947). Energy E in units of t, which is about 2.7 eV, the basic nearest neighbor hopping energy. The Dirac- like features are the linear energy dispersions, present near the neutral points K, K′. One is expanded into the *right panel* of the figure. (Reprinted figure with permission from Castro Neto et al., Fig. 3. Copyright (2009) by the American Physical Society)

The energy near the crossing points is

$$E = \pm\hbar v_F |\mathbf{k} - \mathbf{K}|, \qquad (1.1)$$

where $\mathbf{K}$ (or $\mathbf{K}'$) is the neutral point at the corner of the Brillouin zone. The Fermi velocity determining the slope of the linear sections is given as

$$v_F = 3ta/(2\hbar). \qquad (1.2)$$

This evaluates as 0.91×10^6 m/s, taking the hopping energy t = 2.8 eV, $\hbar = 1.05 \times 10^{-34}$ J-s and nearest neighbor distance a as 142 pm.

The density of states is linear in E and vanishes at E = 0, according to

$$g(E) = 2 A_c |E|/(\pi \hbar^2 v_F^2) \approx g_0 |E|/A_c, \qquad (1.3)$$

where $A_c = 3\sqrt{3} \, a^2/2 = 5.18$ Angstrom^2 is the area of the graphene unit cell and $g_0 = 0.09/(eV^2$ unit cell), with a the interatomic spacing.

Integrating this expression to energy E_F gives the carrier density per unit area as

$$N = E_F^2 g_0/2A_c, \qquad (1.4)$$

so that, e.g., $E_F = 0.28$ eV corresponds to $N = 7 \times 10^{12}$ cm^{-2}. It turns out that these conditions can be met in the common configuration of a graphene monolayer placed above a 300 nm SiO$_2$ layer on doped silicon, if the silicon acting as gate is biased at 100 V. The formula giving the induced charge follows from Gauss's Law of electrostatics in the form

$$N = \varepsilon\varepsilon_0 V_G/de, \tag{1.5}$$

where voltage V_G appears across a dielectric layer of thickness d and permittivity ε, $\varepsilon_0 = 8.85 \times 10^{-12}$ F/m, and electron charge $e = 1.6 \times 10^{-19}$ C. Measuring the electrostatically induced free carrier density N was one of the key initial measurements (via Hall effect) of the pioneering workers Novoselov et al. (2004) and Zhang et al. (2005).

The bands shown in Fig. 1.3 have additional beneficial features that were recognized only well after the work of Wallace. The similar but distinct corner points K and K′ were described by Semenoff (1984), as "right- and left-handed degeneracy points". The distinction relates to the distinct sublattices A and B, and is related to the unusual "two-component wavefunction" for the carriers. While the two components are related to the two sublattices, it is convenient to describe the result as an artificial spin ½.

The desirable aspect of this complexity, as first recognized by Semenoff (1984), is that direct back-scattering of a carrier is forbidden, as it represents a "spin-flip transition", or a transition from A lattice wavefunctions to (orthogonal) B lattice wavefunctions. As was shown by McEuen et al. (1999), the mean free path in metallic nanotubes, whose bandstructure is similar, is exceptionally long on this account. These workers found that the mean free path in metallic carbon nanotubes was about 100 times longer than in similar quality carbon nanotubes that were semiconducting. They attributed this large difference to the change in the band structure from conical in the metallic case to parabolic in the semiconductor case.

The starting point for the two-component wavefunction, valid near $\mathbf{K}$ and $\mathbf{K'}$, is the simplified 2D Dirac Hamiltonian:

$$\hat{H} = b\hbar v_F \boldsymbol{\sigma} \cdot \mathbf{k}, \tag{1.6}$$

where $b = 1 (-1)$ for states of energy above (below) those at point $\mathbf{K}$, and $\boldsymbol{\sigma}$ is a Pauli matrix. This simplification, corresponding to Eq. (1.1) above, comes from cancellation of terms in the usual tight binding theory by application of the $\mathbf{k \cdot p}$ approximation (Semenoff 1984; DiVincenzo and Mele 1984). A review of the algebraic reduction of the Schrodinger theory to the simple Dirac expression (1.5) is given by Castro Neto et al. (2009).

Evidence that graphene is truly a 2D system, that occurs in single layer thickness and indeed can have exceedingly large electron mobility comes from observation of the anomalous Half-integer Quantum Hall Effect with levels given (Novoselov 2011) as

$$E_n = \pm v_F[2e\,\hbar B(n + \tfrac{1}{2} \pm \tfrac{1}{2})]^{1/2} \text{ where } n = 0, 1, 2\ldots \tag{1.7}$$

The details of this effect confirm the unusual band structure and two-component electronic states. The exceptionally high carrier mobility is further indicated by observation of the Quantum Hall Effect at room temperature by Novoselov et al. (2007). No other material has displayed the effect at room temperature.

The mobility was estimated by Novoselov et al. (2007) as ≈ 1 m^2/Vs at room temperature, with an estimate of the scattering time as 10^{-13} s. At 10^6 m/s, the scattering distance is 100 nm, or about 700 bond-lengths. More recent measurements by Mayorov et al. (2011) reveal mobilities ten times higher, ≈ 10 m^2/Vs at room temperature, and confirm earlier reports of ballistic electron transport (Bolotin et al. 2008; Du et al. 2008), now measured at room temperature. The improved mobility values come from samples that are freely suspended and heated to release adsorbed atoms, or mounted on BN, a substrate with fewer charge fluctuations. The record stated mobility now appears to be 100 m^2/Vs, although measured at the low temperature 2 K (Elias et al. 2011).

The conductivity of graphene is dependent on temperature and on the Fermi level position (the doping density). In practice, large doping, achieved by rinsing the graphene in nitric acid (Bae et al. 2010), has led to a resistivity of 30 Ω/square at room temperature in an assembled four-layer graphene sheet. The conductivity of neutral graphene, with the Fermi energy at the Dirac point, where the density of states is zero, has been somewhat puzzling, although not of practical concern as devices will always be designed with enhanced conductivity achieved by doping. While the early literature quoted a minimum conductivity for pure neutral graphene even at low temperature, the theoretically expected Mott-Anderson metal insulator transition for neutral graphene has now been observed by Ponomarenko et al. (2011). The difficulty of observation of the insulator transition has been that in a practical situation stray electric fields from a substrate induce "charge puddles" that make adjoining regions of the putative neutral sample actually conductive, so the true Dirac-point behavior is obscured. The typical observed minimum conductivity is now understood as tunneling of carriers between such randomly distributed positive and negatively doped regions. Only with elaborate electrostatic shielding were these workers able to observe the Mott transition. The experimental confusion has been resolved, but confusion still exists in the theoretical literature. An expected correction, of perhaps 25 %, to the electrical conductivity may arise from electron–electron interactions in graphene, as recently proposed by Rosenstein et al. (2013). These authors discuss "The novelty of the physics of undoped graphene" based on the assumption of a "dc conductivity $\sigma(\omega)/\sigma_0 = 1 + C\ \alpha\ +..$" where $\alpha = e^2/\hbar v$, $C = 0.26$, and $\sigma_0 = e^2/4\hbar$. This assumed value, $\sigma_0 = e^2/4\hbar$, is quite at odds with the quoted work of Ponomarenko et al. (2011), clearly showing that $\sigma_0 = 0$, due to the expected and observed metal–insulator transition in undoped graphene. The reader may be warned that a large body of theory has been based on the erroneous assumption of a metallic groundstate, of conductivity on the order of $e^2/4\hbar$, for neutral graphene.

In practice graphene in monolayer and multilayer forms are superb electrical conductors, typically doped to 10^{12} cm^{-2} by a gate electrode or by chemical doping. For example, exceedingly high values of measured current density have recently been reported by Jain et al. (2012) for graphene on hexagonal BN substrates. The graphene is grown by chemical vapor deposition CVD on Cu foil and then transferred onto mechanically exfoliated flakes of hexagonal BN that, in turn, had been transferred onto an oxidized Si wafer. The electrical characteristics of

Cu-grown CVD graphene on the h-BN surface are excellent, including mobility $\mu = 1.5$ m²/Vs at carrier density 1×10^{12} cm⁻². The breakdown current density was measured as 1.4×10^9 A/cm² for a multilayer on h-BN substrate. This is an exceptionally high value. According to Karim et al. (2009), bulk metals fail because of Joule heating at current densities 10^3–10^4 A/cm², but gold nanowires, of diameter 80 nm withstand 1.2×10^8 A/cm². In comparison, carbon nanotubes, very similar in their conduction to graphene, are reported to withstand current density on the order of 10^9 A/cm² (Dai et al. 1996; Yao et al. 2000). It is clear that the observed values are dependent on the ability of the substrate or environment to remove heat. This aspect likely explains the strong rise in sustainable current density as the diameters of the measured gold nanowires were reduced from 700 to 80 nm. In the graphene case, the thermal conductivity of the substrate seems important, with a note that the same type of CVD-grown graphene multilayer that withstood 1.4×10^9 A/cm² on h-BN withstood only 0.9×10^9 A/cm² when mounted on oxidized silicon, a surface of fused silica.

Optical absorption of graphene occurs by raising the energy of free carriers. Since there is no bandgap, the absorption has no energy threshold, and is broadband in nature. But since the material is only 0.34 nm thick, the absolute absorption of one layer, as suggested earlier, is only about 2.3 % (Nair et al. 2008).

There is no indication in the experimental literature of graphene of any weakness of the crystalline lattice, arising from its 2D nature (A separate question is one of weak nanometer scale out-of-plane ripples or corrugations of the 2D lattice that are sometimes seen. It appears that such small modulations can occur by adsorbed weakly bonded atoms that somewhat alter the strength of in-plane bonds, allowing the deflections especially if the sheet has tension imposed by its boundary). The practical perfection and robust nature of the graphene 2D lattice may be surprising, because there is a respectable literature that says that, in two dimensions (2D), crystals do not exist at finite temperature. More precisely, an *extended* crystalline phase, where each atom can be reached by lattice translations

$$\mathbf{R} = n\mathbf{a} + m\mathbf{b}, \qquad (1.8)$$

with integer n and m, is not possible at finite temperature in 2D. It is not too hard to show (Landau and Lifshitz 1986) that thermal displacements $\mathbf{u}$ of atoms from their lattice positions become indefinitely large in the 2D case of a sample of size L approaching infinity, according to

$$<\mathbf{u}^2> = \text{const.} \times T \ln(L/a). \qquad (1.9)$$

This literature applies to graphene at finite temperature, but its predictions have led to no observed departure of behavior from that of traditional crystals. In part, the corrections may not be observed because the bonding in graphene is so strong, with a Debye temperature above 2,300 K, that the material is usually near the zero temperature limit of its vibrational behavior. Second, the size L of the obtained samples is typically not large, although 30-in. polycrystals have been constructed. Third, the deflections Eq. (1.9) are with respect to a fixed origin, while *inter-atom* vibrational displacements in the 2D case at finite temperature are not anomalous. The motion of near-neighbors is highly correlated, much as in a 3D crystal, but anomalous behavior exists in the relative motion of distant neighbors. In almost all cases the observation apparatus or mounting substrate will act to pin the lattice, so the absolute displacement can be small. Finally, no serious attempt has been made to observe these effects. What is certainly known is that these effects have in no way impeded work on graphene, and samples of graphene have been heated in transmission microscope stages to temperatures above 2,000 °C (Jia et al. 2009) with only small changes, such as reconstruction of the edges, observed to have been caused by the heating. Simulations of graphene using quite realistic bonding potentials (Zakharchenko et al. 2011) have recently suggested that graphene disintegrates into carbon chain fragments around 4,900 K, certainly putting graphene into the class of refractory materials.

1.3   Practical Consequences of One-Atom Thickness

The discussion above, with the exception of the Quantum Hall Effect, is similar to discussion of a three dimensional crystal such as Si. The lattice and band structure calculation method is the same, with the novelty mainly in the conical bands, that do not appear to be a specific consequence of the dimensionality. The surprising and important fact is that the usual breakup of a thin film into islands as its thickness is reduced is avoided entirely by the strong directional 2D bonding in graphene, BN and other members of the class of covalent 2D materials clarified by Geim and Novoselov. So these one-layer materials exist, are actually very robust, and can often be considered largely in the same way as 3D materials are considered.

As mentioned above, in practice graphene is difficult to deal with, because it is nearly invisible and needs to be supported. The softness is a direct consequence of the single atom thickness, and can be discussed classically as mentioned above.

The fundamental oscillation frequency f_0 for singly- and doubly-clamped graphene beams, was experimentally found to be well fitted by the classical expression (Bunch et al. 2007)

$$f_0 = \left\{ \left[A \ (Y/\rho)^{1/2} t/L^2 \right]^2 + A^2 0.57 T/\rho L^2 wt \right\}^{1/2}, \qquad (1.10)$$

where Y is Young's modulus, T the tension applied in Newtons, ρ the mass density; t, w and L are the dimensions of the beam and the clamping coefficient A is 1.03 for a doubly clamped beam and 0.162 for cantilevers. The value of Y was taken as 1 TPa.

Based on such measurements, a spring constant K* (for a doubly-clamped graphene beam of length L, width w and thickness t, with force applied at its center) can be written (Shivaraman et al. 2009) as

$$K^* = 32Ywt^3/L^3. \tag{1.11}$$

As suggested above, because the thickness t entering the formula is only 0.34 nm, this is the smallest possible spring constant K* for a sample of width w and length L. If L is a molecular size, K* is on the scale of tens of Newtons/m, but for L exceeding a micrometer the restoring force F = K*x, for deflection x, is vanishingly small, resulting in a sheet that adheres to any substrate by unavoidable van der Waals attraction. The flexural modes observed by neutron diffraction (Nicklow et al. 1972; see also Mounet and Marzari 2005) on graphite, are quite consistent with the classical analysis outlined here.

The consequence of this softness is that most samples of graphene are supported on a substrate, commonly on SiO_2 grown thermally on P-type silicon. Micrometer scale sheets of un-supported graphene may cross trenches etched in a silicon surface or cross openings in TEM grids, but in device applications the graphene must lie on a rigid support. The support may distort the graphene from local microscopic planarity if it is itself imperfect, as in the case of amorphous silica. The support may contain electrical charges that can induce charges in the graphene, that will alter its local Fermi energy. Careful research reveals (Dean et al. 2010; Xue et al. 2011) that non-polar crystalline support surfaces, such as boron nitride BN, yield much smoother graphene surfaces, as seen by scanning tunneling microscope, than graphene on silica.

While the main interest and novelty is in monolayer graphene, *bilayer graphene* can be grown under similar conditions as used for the monolayer, for example see Ohta et al. (2006). In its characteristic form, with Bernal stacking, carbon atoms on the upper layer are centered on rings (holes) in the lower layer. This registry leads to a semimetal with conventional parabolic bands and massive carriers, lacking the conical band structure. Bilayer graphene retains the high carrier mobility because of the long range perfection of covalent bonding, but the formal cancellation of direct backscattering not longer appears. (Bilayer graphene does not usually result from stacking of graphene monolayers, in which case there is typically no particular registry of angles between the layers. In the case of stacked graphene monolayers the transverse conductivity is additive as in a parallel conductor arrangement, but the lack of registry means that the conduction perpendicular to the planes is reduced.) In the chemical methods of breaking up graphite into flakes, it is common to find stacks of variable numbers of monolayers. Most chemical methods actually produce graphite oxide, an insulator, that is later reduced back to an oxide-free material, however, commonly retaining many defects in the planes that quite substantially reduce the electrical conductivity.

The softness and related aspects of graphene are as one would expect for a material of vanishing thickness, a 2D material. A fundamental problem in practice is growing planar crystals of graphene. This central problem will be discussed in detail in Chap. 2, but we offer below a brief discussion. On a more detailed level, the graphene crystal lattice vibrations differ from a typical 3D crystal because it can flex perpendicular to its plane. Actually, however, the flexural modes of graphene are almost unchanged from the flexural modes that are well-studied in

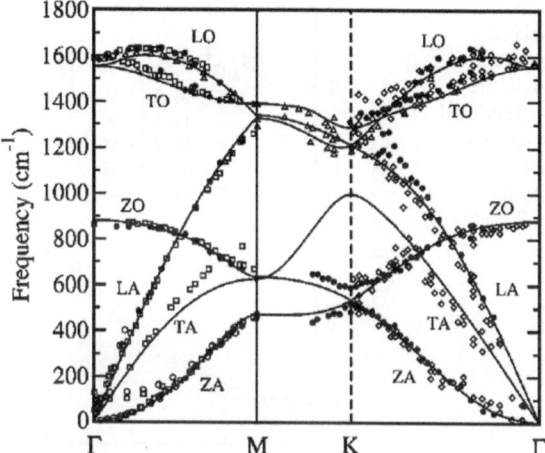

Fig. 1.4 Vibrational frequencies calculated for graphene in the ab initio method of Mounet and Marzari (2005), compared with data points obtained experimentally for graphite. Here the horizontal axis is wavevector, with the Brillouin zone points Γ, M and K as defined in Fig. 1.2. The modes denoted ZA are the flexural motions, described in the text, that appear in 2D layers. (Reprinted figure with permission from Mounet and Marzari, Fig. 5. Copyright (2005) by the American Physical Society)

graphite. Since graphite is only van der Waals bonded in the perpendicular direction, its lattice modes include the flexural motions.

The modes labeled ZA in Fig. 1.4 have a quadratic dependence of frequency on wavevector, characteristic of flexural motions of a 2D layer in 3D space. That these flexural modes are observed in three-dimensional graphite is an indication of the weak coupling of the graphene layers by van der Waals forces in graphite. An evidence of these modes is the negative temperature coefficient of in-plane expansion, observed in graphite upto 600 K, and predicted to a much higher temperature in graphene. The peak frequencies in the spectrum are near 1,600 cm^{-1}, that, with the conversion 1 meV = 8.065 cm^{-1}, corresponds to 198.4 meV. The very high vibrational frequencies are an indication of the strong covalent bonding and small atomic mass in graphene, and these factors correlate with high thermal conductivity. The high frequency optical modes are less likely to be excited by carrier motion, a fact that contributes to the high carrier mobility in graphene.

Extremely high current densities in graphene, on the order of several mA/μm, are reported, for example, in the measurements of Liao et al. (2010). Assuming the thickness of the graphene is 0.34 nm, the current density is approximately 3×10^8 A/cm^2.

A thorough experimental and theoretical analysis of the phonon-induced resistivity of graphene is given by Efetov and Kim (2010). A key result of their work is illustrated in Fig. 1.5. The experiment achieved exceptionally large carrier

Fig. 1.5 Normalized
dependence of graphene
resistivity on temperature,
dominated by 2D acoustic
phonon scattering. (Reprinted
figure with permission from
Efetov and Kim, Fig. 4.
Copyright (2010) by the
American Physical Society)

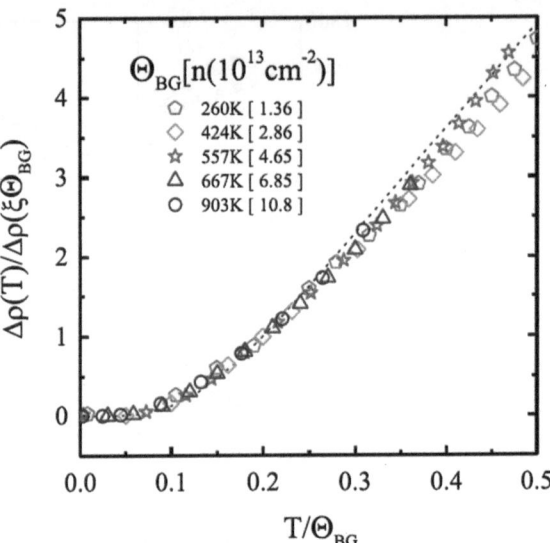

densities N by the electric field effect applied by a "solid-electrolyte polymer gate", as described in their paper. By this means they achieved a specific capacitance between graphene and the gate, in which Li^+ and ClO_4^- ions are mobile in the polymer PEO poly(ethylene)oxide, of 3.2 μF/cm^2, more than 250 times larger than the capacitance available in the conventional P-Si/SiO$_2$ gate, described in Eq. (1.5). The large induced mobile carrier concentrations, N, were measured by Hall effect. The conventional understanding of the temperature dependence of resistivity in metals is based on the Debye temperature, that is dependent on the lattice vibration energies (such as shown in Fig. 1.4 for graphene). According to Krumhansl and Brooks (1953), the atomic motions of carbon atoms in graphite can be described by Debye temperatures $\theta_D \approx 2{,}500$ K for in-plane motions and $\theta_D \approx 900$ K for out of plane motions. This would suggest a Debye temperature for graphene of 2,500 K, while Efetov and Kim quote 2,300 K, in reasonable agreement. Efetov and Kim adapt the resistivity analysis from a 3D metal to the 2D system where the carrier density is widely variable (in their work via the voltage on the "solid-electrolyte polymer gate"). They find the effective Debye temperature, now called the Bloch-Gruneisen temperature

$$\theta_{BG} = 2\hbar v_S k_F/k_B < \theta_D, \tag{1.12}$$

that marks the change in the temperature dependence of the resistivity, $\rho(T)$, from linear in T to linear in T^4. This parameter, θ_{BG}, is dependent on carrier density, through the density dependence of the Fermi wavevector $k_F \propto \sqrt{N}$. In Eq. (1.12), v_S and k_B, respectively, are the speed of sound and Boltzmann's constant $k_B = 1.38 \times 10^{-23}$ J/K. As shown in Fig. 1.5, the fitted values of the Bloch-Gruneisen temperature $\theta_{BG} \propto \sqrt{N}$ vary from 260 K at 13.6×10^{12}/cm^2 to 903 K at

10.8×10^{13}/cm^2. These authors achieve a universal scaling of the normalized resistivity of graphene, $\Delta\rho(T)/\Delta\rho(\xi\theta_{BG})$, over the whole concentration range, as a function of the normalized temperature T/θ_{BG}. The dashed curve in Fig. 1.5, that quite nicely fits the wide range of experimental data points, represents the theory of Efetov and Kim, with choice of the single parameter $\xi = 0.2$, as well as values $v_F = 10^6$ m/s, $v_S = (2.6 \pm 0.4) \times 10^4$ m/s and an acoustic phonon deformation potential value $D_A = (25 \pm 5)$ eV, all reasonable values. The theoretical fit has included the unusual absence of direct backscattering, as one of the changes from the conventional 3D analysis made by the authors. It is seen that the resistivity at the lowest temperature is residual, depending on impurity and defect levels; rises initially as T^4 and then shifts, above the concentration-dependent characteristic temperature $\theta_{BG} = 2\hbar v_S k_F/k_B$, to a concentration-independent T-linear resistivity.

The resistivity analysis given above applies at small current density j = Neu, with u the drift velocity. In device applications at high current density, velocity saturation is observed. The Fermi velocity in graphene has been mentioned above as near 10^6 m/s. The achieved drift velocity, u, smaller than the Fermi velocity, is an important parameter in connection with device performance, as for example, in the channel of a field effect transistor. The carrier drift velocity u can be extracted from I–V measurements, such that I = Wj, for a channel of width W, using the relation j = Neu.

DaSilva et al. (2010) have studied the carrier drift velocity and its saturation in graphene. They find that u $\approx$ 0.3 v_F, at electric field E $\sim$ 0.6 V/μm, for mechanically exfoliated graphene placed on the traditional P–Si/SiO$_2$ substrate. In their analysis of high-electric-field data, they find that scattering by graphene phonons (as described by Efetov and Kim) is insufficient to explain their measurements. DaSilva et al. find that the primary scattering mechanism at high current density is emission of surface optical (SO) phonons of the underlying SiO$_2$ substrate. It appears that one of the advantages of the diamond-like-carbon substrate, DLC, is that its phonons are at a higher energy, 165 meV, and less likely to be excited than the phonons in the more common substrate, oxidized silicon.

A property of graphene, closely related to its phonon spectrum, is the Raman spectrum. In Raman scattering, a photon falling on the sample, can be re-emitted, after losing a characteristic excitation energy to the sample. Experimental apparatus filters out the large number of elastically scattered photons re-emitted at the same frequency, to measure the spectrum of light whose photon energy loss is in a range up to perhaps 0.4 eV (3226 cm^{-1}). The Raman spectra of graphene and graphite are similar, as one might expect, but careful observation reveals small characteristic shifts in the Raman lines, going from single-layer graphene to multiple layers, as illustrated in Fig. 1.6 (Ferrari et al. 2006). The peak near 2,700 cm^{-1} ($\sim$8,065 cm^{-1}/eV), known as the "2D peak", arises from phonons near the K point in the Brillouin zone (See Fig. 1.4). A photon from the 514 nm laser is absorbed and re-emitted with a shift of about two phonon energies, in the Raman process for single layer graphene shown in Fig. 1.6 (Ferrari et al. 2006). These energies represent characteristic "molecular" vibrations of 6-fold carbon, for example, a breathing mode for the benzene ring. The phonon energy is then

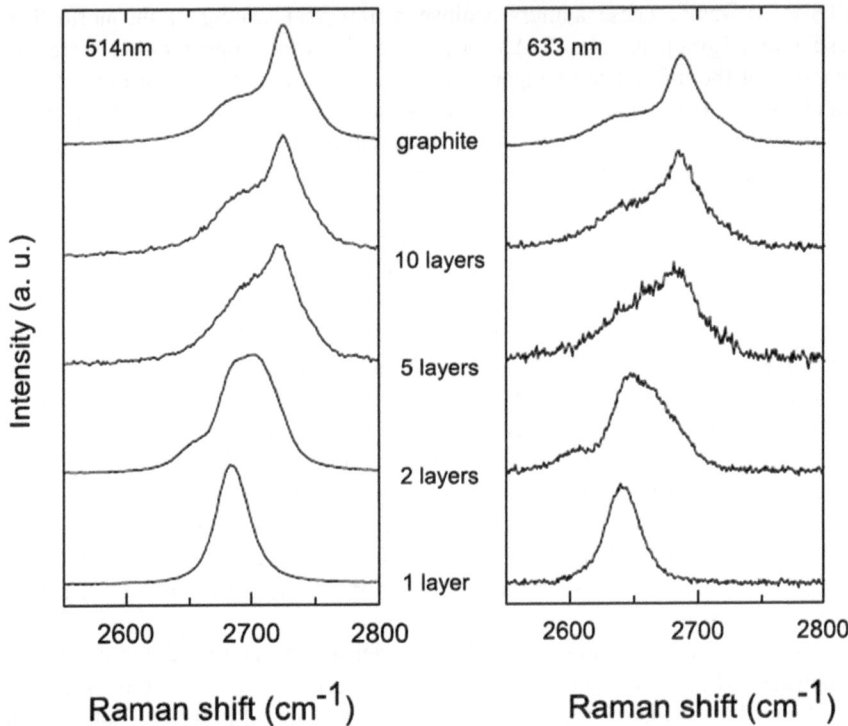

Fig. 1.6 Demonstration of small changes with graphene layer thickness on the *Raman lines* observed at 514 and 623 nm, *left* to *right*. (Reprinted figure with permission from Ferrari et al., Fig. 2 (b, c). Copyright (2006) by the American Physical Society)

about $1,350 \text{ cm}^{-1} = 2,700/[2 \times 8,065] = 0.167$ eV. This energy is close to the observed value near the K point in the neutron spectroscopy of Nicklow et al. (1972). Data points from this work are plotted in Fig. 1.4, along with the ab initio theory of Mounet and Marzari (2005) (solid curves). The details of the structural shifts in the spectra with layer thickness are explained in some detail by Ferrari et al. (2006).

These authors conclude that graphene's electronic structure is revealed by the Raman spectrum, that clearly evolves with the number of layers. Raman finger-prints for single, double and few-layer graphene reflect changes in the electronic structure and electron–phonon interactions, to allow unambiguous, nondestructive identification of graphene layers. The authors point out that in AFM or TEM measurements it is, in practice, only possible to distinguish between one- and two-layer graphene if the films contain folds or wrinkles. A review of Raman work on graphene is given by Malard et al. (2009).

Another phonon-related physical property is the thermal conductivity. Because of the high phonon energy typical of graphene, one expects high thermal conductivity. The thermal conductivity, κ, of supported graphene has been studied by Seol et al. (2010). Measurements on supported graphene reveal values of κ near 600 W/(m-K). This value, from graphene on the traditional SiO_2 on Si substrate, exceeds that of common electronic materials, such as copper. This value $\kappa = 600$ W/m-K, however, is less than values, in the range 4,840–5,300 W/(m-K), inferred (Balandin et al. 2008) from Raman measurements on *suspended* graphene, and 3,500 W/m-K on single-wall carbon nanotubes (Pop et al. 2006).

These large thermal conductivity values, exceeding those of diamond and graphite, are consequences of the strong bonding of the light carbon atoms. The reduction of thermal conductivity, for samples mounted on fused quartz, is attributed to phonon-substrate scattering. In more detail, phonons leak from the graphene into the substrate, and the flexural modes, that are predicted to contribute strongly to the phonon conductivity, are strongly scattered by the substrate.

In discussing matters of practical importance for applications of graphene, it is fortunate that one can largely dismiss practical consequences of the esoteric effect known as Klein tunneling. This process was predicted by Klein (1929) and reanalyzed by Katsnelson et al. (2006) in the context of graphene. The effect arises in pnp or npn junctions of graphene as a consequence of the symmetry of the gapless conical bands about the neutral point $E = 0$. For example, in the npn case, with E_F at a positive value on the outer electrodes, transport can occur with no reduction, in the p-region of width D, at its symmetrically located Fermi level, $-E_F$, but raised in potential by $V_o \approx 2\,E_F$. Formally this can be regarded as a hole perfectly carrying the current through the barrier.

The electron of energy E approaches a barrier of height V_0, in the center, of width D, where the potential is raised. The electron on the left, with positive wavevector k and positive pseudospin ($\rightarrow$), matches, in energy and pseudospin, a positive hole of opposite wavevector q. The current is thus continuous in the barrier, carried by the matching hole. The same transformation occurs between the barrier and the right hand electrode. This gives unit probability of transfer (at precise normal incidence) of the electron by Klein tunneling from left to right. As long as the interfaces are sharp and the barrier permits coherent ballistic motion, there is no limitation of unit transmission from increasing width and/or height of the barrier at precisely normal incidence, $\alpha = 0$.

The corresponding transmission factor $T = 1 - |r^2| = |t^2|$, in the limit of a high barrier $|V_0| \gg |E|$ is

$$T = |\,t^2\,| = \cos^2(\alpha)/[1 - \cos^2(q_x D)\sin^2(\alpha)]. \qquad (1.13)$$

These two equations also show that perfect transmission for $\alpha = 0$ reappears periodically at other angles, specified by

$$q_x D = n\,\pi, \quad \text{where } n = 0, \pm 1, \pm 2, , \ldots \qquad (1.14)$$

In these equations

$$q_x = \left[(E - V_0)^2/\hbar^2 v_F^2 - k_y^2\right]^{1/2}. \tag{1.15}$$

It appears that, under a more realistic analysis, the individual pn junctions in the graphene further reduce the transmission, for angles away from precise normal incidence. This was analyzed carefully by Cheianov and Fal'ko (2006), who derive the transmission probability T of the single pn junction as

$$T = \exp\left[-\pi(k_F D)\sin^2(\alpha)\right] = \exp\left[-\pi \hbar v_F k_y^2/(eE)\right], \tag{1.16}$$

with E the electric field at the junction. The second form of the equation is presented by Young and Kim (2009) and attributed to Cheianov and Fal'ko (2006). Strong collimation, $T = 1$, is retained at $\alpha = 0$, but now transmission off normal incidence, $k_y^2 > 0$, is greatly reduced by the traditional exponentially decaying tunneling barrier transmission factor.

The details of the Klein tunneling effect in graphene have been confirmed in careful experimental work by Young and Kim (2009), but the effects in the end are minor and indeed hard to observe. One can conclude confidently that the effect is not of importance in the applications of graphene.

Applications of graphene in some cases are hindered by the lack of an energy gap. Graphene Nanoribbons (GNR) and nanomeshes are, respectively, ribbons or periodically perforated sheets of graphene, that do develop bandgaps, up to perhaps 0.2 eV. Nanoribbons of width W are found to have energy gaps arising from the boundary condition. In detail the energies available depend upon the precise boundary condition, but an empirical rule is E_G (eV) $\approx$ 2 eV nm/W(nm). Figure 1.7 shows a TEM image with examples of zigzag and armchair edges of graphene, note the scale bar of 1 nm. The smooth edges were obtained by heating the graphene above 2000 K. The zigzag and armchair edges, respectively, can be visualized also in Fig. 1.1 as the vertical and horizontal boundaries of the Figure.

Confinement in a ribbon in general leads to an energy gap inversely related to the width W. The propagating wavefunction along a ribbon is

$$\psi(x, y) \sim e^{ikx}\sin(n\pi y/W), \tag{1.17}$$

where the 2nd factor recognizes that the probability of finding the particle at $y = 0$ or $y = W$ must vanish, as a work function barrier confines the particle. This gives $k_y = n\pi/W$, that corresponds, in a conventional metal (or in *bilayer* graphene, with massive electrons moving in parabolic bands), to a confinement energy

$$(\hbar k_y)^2/2m = h^2/8mW^2. \tag{1.18).}$$

In monolayer graphene, however, as was pointed out by Berger et al. (2006), the linear bands lead to a different GNR energy gap. In monolayer graphene the energy E_n at the bottom of nth sub band state becomes, with v_F the Fermi velocity,

Fig. 1.7 Graphene annealed under high electrical current density shows well-defined edges, as marked. 1nm scale bar. (Reprinted figure with permission from Jia et al., Fig. 1f. Copyright (2009) by the Science)

$$E_n = v_F|p| = \hbar v_F\left(k_x^2 + k_y^2\right)^{1/2} = \left[E_x^2 + n^2(\Delta E)^2\right]^{1/2} \qquad (1.19)$$

where n = 1, 2, 3,...

$$\Delta E = \pi \hbar\, v_F/W \sim 2\ eV\ nm/W, \qquad (1.20)$$

and $E_x = \hbar v_F k_x$.

This gives a gap ~ 100 meV for nanoribbon width W = 20 nm, quite close to the measurement of Han et al. (2007).

To this basic analysis one must add consideration of the nature of the edges. In practice it is difficult to achieve a uniform boundary, and the size- and boundary-effects are obscured if the material has a short mean free path and if the boundaries are rough.

If a bandgap of 0.4 eV is needed for a conventional transistor, this would require a width W = 5 nm, approximately the height of Fig. 1.7, that is difficult to manufacture with specific boundary type. In spite of this difficulty, an analysis of graphene nanoribbon field effect transistors using boron-doped material is given by Marconcini et al. (2012).

The resulting transistor will carry a rather small current in its ON condition, because the channel cross-section is so small. The measured currents in graphene are at most 1 mA/μm, that would correspond to at most 200 nA at W = 5 nm. An alternative approach is the *graphene nanomesh* (GNM), graphene with a regular pattern of perforations. Two recent papers (Bai et al. 2010; Liang et al. 2010) have found methods of making graphene nanomesh based on self-organizing domains within a block co-polymer to provide an etching mask. The actual etching of the graphene is done with reactive ions, leading to patterns of holes with periodicity on the order of 40 nm, with minimum "neck widths" that play the role of ribbon width, W, variable from 7 to 15 nm. The overall larger cross-section of the graphene nanomesh GNM allows a larger total ON current, the device may be regarded as a parallel array of nanoribbons. Both groups have demonstrated transistors of the FET type using the nanomesh layers, but this work is in its infancy.

In conclusion, the transport properties of graphene are extremely favorable, and attainable at the limit of one atom thickness. A main practical difficulty, to exploit the superior properties, lies in preparing large high quality samples. One important consequence of the one-atom thickness is that one cannot grow the crystal layer by layer on a substrate, because it is only one layer in toto. (Such a growth may apply to crystalline graphite in its geological origins, but removing the single layer from graphite, as was accomplished in a reliable but painstaking fashion by Novoselov et al. (2004), is not a practical process.) While a carbon nanotube will grow vertically, under CVD conditions, from a single transition metal catalyst particle of diameter near 1 nm, an analogous growth of graphene, perhaps from a line of catalysts, on a planar surface, has not been accomplished. On the transition metal surface, commonly Ni or Cu, CVD growth of graphene is of islands nucleated at random sites, that coalesce to form 2D polycrystals. Thus, one finds arrays of 2D crystalline grains connected by boundaries, with no gaps between grains. A careful study by Kim et al. (2011), of single layer graphene grown on Cu foil of 100 μm grain-size, shows well-connected grains of 5–7 μm size, with a distribution of mismatch angles between grains peaking near 0° and near 30°. The authors suggest that this indicates a tendency for nucleated graphene islands (grains of a 2D lattice) to align with the lattice of the large underlying Cu grain. It appears that the grain size of graphene grown by CVD is larger in case of slower growth, as discussed by Tsen et al. (2012), and not closely related to grain size of the underlying Cu catalytic foil. To obtain the single graphene layer, the usual practice is to dissolve the Cu foil chemically.

In their study of high quality catalytically grown graphene polycrystals, Kim et al. (2011) find that gaps between grains do not exist, and, for example, large angle grain boundaries are accomplished by reconstructions, such as alternating 5- and 6-fold rings along the grain boundary. The strength of the sp^2 planar bonds, even if somewhat distorted, allows the graphene film, grown at 1,000 °C, to be continuous, to avoid gaps between grains of different planar orientation, and to grow over local imperfections including steps in the substrate. The results are high quality 2D polycrystals, that have been achieved in lateral dimension nearly 1 m, in the work of Bae et al. (2010). As mentioned, Bunch et al. (2008) found that graphene, as peeled mechanically from graphite (likely a single crystal with no internal boundaries) is impermeable to any gas, including helium. It seems reasonable to suppose that the same impermeability applies for the catalytically-grown polycrystalline films, since the grain boundaries adjust local bonding to avoid gaps.

Chapter 2
Practical Productions of Graphene, Supply and Cost

Graphene in small platelets of high aspect ratio is available by chemical exfoliation of graphite, with a variety of methods and a long history. Graphite is a widely available mineral, at a cost between $1.50 and $2.00 per kg, according to "Mineral Commodity Summaries 2013" published by US Department of the Interior and US Geological Survey. (For comparison, the same source estimates indium metal at ~ $650 /kg, with a separate estimate for indium tin oxide, ITO, that is deposited on glass to make the leading form of transparent conductor, as $800 /kg.) Typically, however, the exfoliation processes leave nano-platelets of only 100–500 nm lateral extent, single monolayer to tens of monolayers in thickness, with defects primarily, but not entirely, at the edges. A contemporary example is "Graphene Nanoplatelets, Grade C" sold by XG Sciences, of Lansing MI, USA. This product, that can be specified at surface area 300, 500 or 750 m^2/g, is described as "aggregates of sub-micron platelets that have a particle diameter less than 2 microns and particle thickness of a few nanometers, depending on the surface area". The XG Sciences data sheet, available at http://xgsciences.com/products/graphene-nanoplatelets/grade-c/, indicates a substantial impurity level, 10 % by weight of oxygen, with the comment "Nanoplatelets have naturally occurring functional groups like ethers, carbonyls, or hydroxyls....present on the edges of the particles and their wt % varies with particle size." This may be expected in any graphene platelet, since unfilled trigonal carbon bonds at the edges will easily react with ambient gas. This may not be serious, because the impurity will be weakly held and probably can be driven off with annealing, perhaps followed hydrogen annealing to fill the bonds, at the end of processing. While a price list is not available, Samba Sivudu and Mahajan (2012) suggest a cost in excess of $220 /kg for a similar XG Sciences product. (For comparison, Norit GSX, an "activated carbon" characterized by surface area about 950 m^2/g, with curved rather than flat morphology, and possessing lateral planar extents less than 20 nm, is offered by Alfa Aesar at $290 for 2 kg. Such "activated charcoal" products are sold in large volume for water purification.) To make a coherent electrical conductor, platelets such as offered by XG Sciences must be deposited, typically by spinning, to produce a layer whose conductivity depends on the contact between the overlapping individual grains, that are too refractory to sinter. Other important forms of small particle carbon come from heating

E. L. Wolf, *Applications of Graphene*, SpringerBriefs in Materials, DOI: 10.1007/978-3-319-03946-6_2, © Edward L. Wolf 2014

petroleum, coal, peat, coconut shells, or chemicals such as polyacrylonitrile (PAN, rayon) in the absence of oxygen. These typically are large volume, low cost, processes. One practical distinction between exfoliated graphene platelets and the other forms of particulate carbon is the large aspect ratio of the platelets.

Centimeter- and even meter- scale sheets of monolayer graphene are a recent development, presently largely a laboratory phenomenon, but clearly destined for electronic applications. The primary source of large sheets of monolayer graphene is by chemical vapor deposition of a hydrocarbon such as methane (or benzene) onto a hot Cu or Ni surface. A second source is by heating single crystals of SiC such that Si sublimates and C precipitates as graphene in an epitaxial fashion. In the first case the copper substrate is usually not useful and is removed by its chemical dissolution. As mentioned earlier, Bae et al. (2010) have achieved polycrystals of lateral extent one meter by such a method, finding a practical approach to removing the copper substrate. That this is not an inexpensive process is suggested by the selling price (Graphene Square, Seoul, South Korea) of $264 for a single 50×50 mm sheet of monolayer graphene on copper foil.

2.1 Graphite-Based Methods

Graphene, in multi-layer, nano-platelet forms, has long been available inexpensively by chemical "exfoliation" treatments of graphite, a mineral found in the earth. These forms are generally used as additives to composites, bringing electrical conductivity and mechanical flexibility and strength. It is unclear at the moment whether such graphite-based materials will be of sufficiently high conductivity, when formed into electrodes, to suffice in applications such as solar cells or interconnects in chip manufacture. An assessment of production of higher quality exfoliated graphene, on the scale of tons per year, was given by Segal (2009). A recent excellent survey of the whole spectrum of graphene applications has been given by Novoselov et al. (2012), who are able to project costs for the various types of graphene and to predict where opportunities for market penetration will appear.

The chemical exfoliation of graphite, first reported by Brodie (1859) and Staudenmaier (1898), proceeds by treating graphite with acid leading to graphite oxide flakes (originally called "graphon"). Graphite oxide is an insulating layer compound with interplane spacing approximately 0.65–0.75 nm. According to Novoselov (2011), graphite oxide can be regarded as graphite intercalated with oxygen and hydroxyl groups, thus a hydrophilic material easily dispersed in water. Such a dispersion includes extremely thin, even monolayer, oxide flakes that can be subsequently reduced, e.g., using hydrazine, according to Ruess and Vogt (1948) and Hummers et al. (1958), or by rapid heating in inert gas (Schniepp et al. 2006) to pure carbon. Such samples, as mentioned, are now regarded as low-quality graphene, although it is clear that the low cost may place them in applications. A suggested structure for Graphite Oxide, with prominent sp^3 bonding of carbon and oxygen atoms (due to Jo et al. 2012) is shown in Fig. 2.1. This Figure also schematically shows the reduced defective graphene obtained with hydrazine treatment. Quoting Mkhoyan et al. (2009), "the graphite oxide is rough, with an average surface roughness 0.6 nm and the structure is predominantly amorphous due to distortions from sp^3 C–O bonds. Around 40 % sp^3 (tetrahedral) bonding was found..". The nature of the "low quality graphene" following reduction of graphite oxide was also characterized in a careful and extensive study of Schniepp et al. (2006).

In their experimental work, Schniepp et al. start with graphite flakes that are reacted in an oxidizing solution of sulfuric acid, nitric acid, and potassium chlorate (similar to the above-cited process of 1898). Schniepp et al. (2006) find that 96 h in the oxidizing solution is needed to completely remove, in x-ray examination, the 0.34 nm interplanar spacing characteristic of graphite, to supplant that spacing with a 0.65–0.75 nm spacing characteristic of graphite oxide in its solid phase. The treatment is likely to form OH hydroxyl, C-O-C epoxide and COOH (O=C-OH) carboxyl groups, mostly near defect sites. The following thermal exfoliation of the solid graphite oxide flakes is akin to an explosion releasing CO_2 gas. This is accomplished by placing a sample of completely dried solid-phase graphite oxide, in a quartz tube purged with argon gas, into a furnace preheated to 1,050 °C. The rapid (>2,000 °C/min) heating splits the graphite into single sheets through the evolution of CO_2 gas. Atomic Force microscopy was among the methods used to establish the single layer nature of the resulting graphene particles, whose lateral dimension is typically 200–500 nm. This careful modern study of the early (Brodie 1859) inexpensive, bulk chemical exfoliation process, and its several variations, concludes that single-layer graphene can (but rarely does) result from such processes, but that any such graphene retains in-plane defects, mainly structural defects, such as kinks and 5-8-5 defects remnant of the successive oxidizing and reduction reactions. (A 5-8-5 defect replaces three 6-membered rings with rings containing 5 and 8 bonds.) In addition one finds remnant chemical impurities like C-O-C (epoxy) or C-OH groups (within the graphene planes) and C-OH and –COOH groups at the edges. The defects clearly remain in the basal plane of the final graphene, as well as at its edges. The defects may have a useful functionalizing role in an additive to a composite material, but are definitely undesirable in graphene intended as a conductive electronic component. Schniepp et al. (2006) measure the electrical conductivity of compacts of their powdered material in the range 1,000–2,300 S/m (0.043–0.1 Ω-cm). These resistivities, somewhat smaller than for other small-particle carbon products, are probably still dominated by the nature of interparticle contacts.

A device-oriented study of nanoribbons produced by chemical exfoliation is that of Li et al. (2008). These authors find chemical means of producing narrow nanoribbons with smooth boundaries starting with graphite oxide, and go on to closely characterize the nanoribbons by electrical methods. The method of Li et al. (2008) is based on a commercial product, "expandable graphite" (Grafguard 160-50 N, Graftech Incorporated, Cleveland, OH). It appears that this material is graphite oxide, since it is readily expanded by heating, as Li et al. (2008) have

(c)

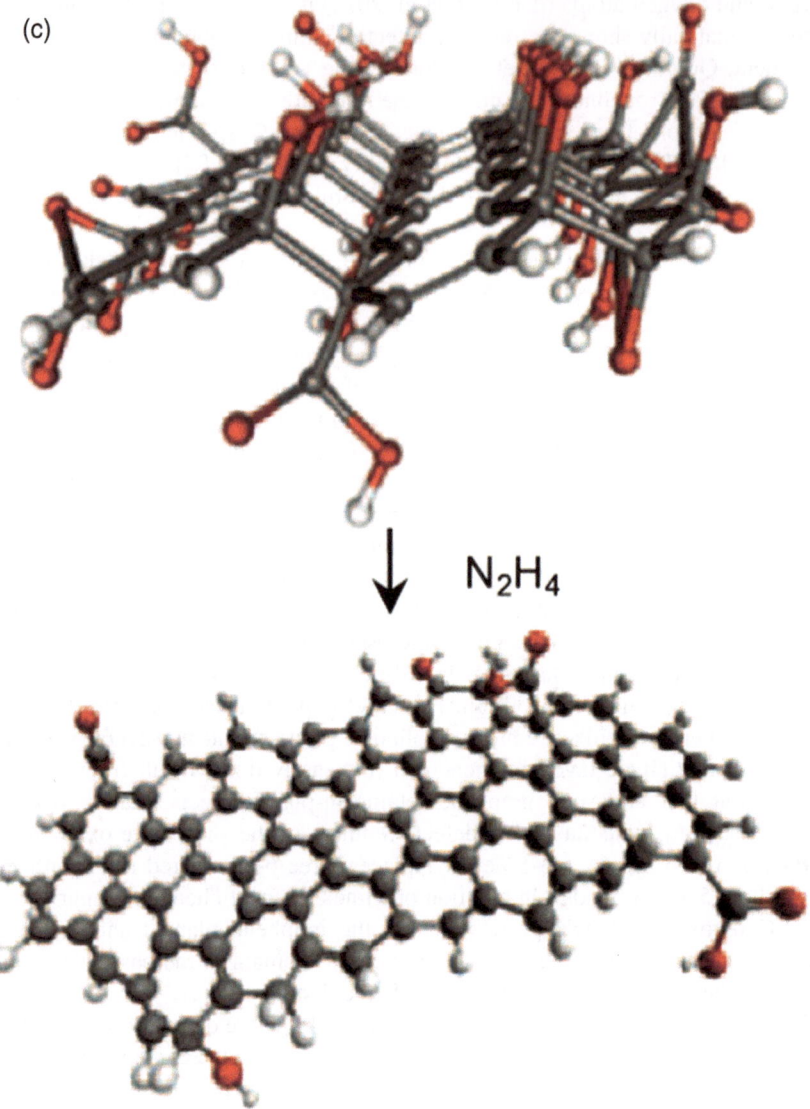

N_2H_4

Fig. 2.1 Schematic diagram of reduction (using hydrazine N_2H_4) of graphite oxide to graphene with retained defects. *Open symbols* are hydrogen, others are carbon and oxygen. (Jo et al. 2012, Fig. 2c) (Reprinted figure with permission from Tung et al., Fig. 1. Copyright (2009) by the Nature Nanotechnology)

done. The "expandable graphite" is exfoliated by heating for 60 s at 1,000 °C in forming gas (3 % hydrogen in argon). Volatile gaseous species violently form from the intercalants, and exfoliate the material into a loose stack of few-layered graphene sheets. The thermal exfoliation step is critical and responsible for the formation of one- to few-layer graphene, and was revealed to Li et al. (2008) by a volume expansion by ~ 100–200 times after exfoliation.

The resulting exfoliated graphite was dispersed in a 1,2-dichloroethane (DCE) solution of poly(m-phenylenevinylene-co-2,5-dioctoxy-p-phenylenevinylene) (PmPV), by sonication for about 30 min to form a homogeneous suspension. (Centrifugation then removed large pieces and the remaining suspension was surveyed for its content of planes and ribbons of graphene.) The PmPV polymer non-covalently functionalizes the exfoliated graphene, leading to a homogeneous black suspension during the sonication process. The authors note that the PmPV was necessary to reach a stable suspension. It appears that the polymer adheres to graphene by van der Waals forces similar in magnitude to the binding force of graphene in graphite itself, and that otherwise graphene will not disperse even in the organic solvent.

Li et al. (2008) found that the sonication time to produce ribbons should be optimized, since ribbons were no longer found after hours of sonication. Extended sonication evidently breaks up ribbons and leads to ever-smaller particles. In general, the GNR (graphene nano ribbon) content was smaller than that of sheets: the solution after centrifugation contains micrometer-sized graphene sheets and nanoribbons. The survey of the solution products was carried out using atomic force microscopy (AFM), after removing the PmPV by calcining at 400 °C. Particular interest was in finding nanoribbons in the sonicated solution products, that were tested by forming a device geometry. Thus, the nanoribbbbons, harvested from the sonicated suspension as described above, were built into field-effect transistor-like structures, choosing ribbons with widths W in the range 10–55 nm. These ribbons were placed on an oxidized p^{++} Si wafer, that formed the back-gate of the field-effect transistor (FET). Palladium Pd contacts were attached to the nanoribbons, to act as source and drain. The devices were surveyed as to the On/Off current ratios, that were found to rise dramatically at small nanoribbon widths W, less than 10 nm. More details on similar devices regarded as sub-10 nm graphene nanoribbon GNR field- effect transistors were given by Wang et al. (2008). The study indicated that an energy gap dependent on ribbon width is a reliable feature. Although these ribbons must have defects in the graphene planes as a consequence of the chemical oxidation/reduction sequence employed, the main impediment to their use in electronics is similar to that of carbon nanotubes, regarding the difficulty in placing many similar elements in exact locations on a chip.

A proprietary chemical approach leads to the well-known commercial product Grafoil, where graphite oxide is reduced back to pure carbon in thicknesses of hundreds of layers. Grafoil is in the form of a cloth used for sealing joints at high temperature and as a moderator of high energy neutrons in nuclear reactors, because of the low mass of the carbon atom. In the paper of Schiffer et al. (1993), the surface area of Grafoil was measured to be 33.3 m^2 for 2.55 g. (The surface area of fully exfoliated graphite is much larger, namely 2,630 m^2/g) This surface area suggests sheet thickness of hundreds of graphene layers. The "Grafoil GTA

Premium Flexible Graphite" sheet has a density of 1.12 g/cc (compared with 2.27 g/cc for single crystal graphite). It is available in a variety of thicknesses.

For the higher quality electronic applications that may include solar cell electrodes and chip interconnects, variants on the exfoliation process that lead to higher electrical conductivity, but still low cost, are needed. The basic difficulty is that graphite flakes are hydrophobic and will not disperse in water under sonication, and the chemical method of oxidation followed by reduction leads to graphene planes with many defects and lower inherent conductivity.

We mention three alternative approaches to exfoliation of graphite, avoiding the acid oxidation followed by reduction. These are: intercalation with alkali metals, direct exfoliation by sonication in organic solvents, and a process called "edge-carboxylation". In these processes, no damage is done to the inner portions of the graphene planes, and the exfoliation occurs from the edges of the graphite flakes without oxidation of the graphene planes.

In the intercalation-based exfoliation method of Shioyama (2001) the graphite was first intercalated with potassium, by heating in the presence of K vapor, at a temperature as low as 200 °C. This leads to electron doping of the graphene, and stoichiometric intercalation phases of graphite are well known (Dresselhaus and Dresselhaus 1981).

Shioyama found that heating the resulting K-Graphite Intercalation Compound (KC_8) at room temperature in a pressure 67 kPa of the vapor 1,3-butadiene (with similar results for the vapor of styrene) led to an exfoliation reaction, evident by the expansion of the K-GIC along its c-axis. The reaction proceeded until all the butadiene vapor was used up. Shiyoama suggested that linear polymers were growing, forcing the graphite planes apart. Heating the resulting black elastic polymer above 400 °C resulted in complete release of the potassium polymers, leaving a residue of pure graphitic carbon, completely exfoliated. This process leaves the graphene planes intact with high electrical conductivity. The refractory and inert nature of graphene aids in this process, but high temperature processes are often difficult and expensive from a production viewpoint.

Classes of organic solvents can disperse graphite flakes under sonication. Liquid phase exfoliation without the oxidation/reduction steps used by Schniepp et al. (2006), and earlier workers (thus, strictly speaking, not a *chemical* exfoliation, but by directly dispersing graphite in other, mostly organic, solvents), has been discussed by Hernandez et al. (2008). Their method starts with powdered graphite and leads to dispersions of graphene in the solvent at concentrations up to 0.01 mg/ml. In the case of solvent NMP (N-methylpyrrolidone) a plot of the thickness of the dispersed flakes peaks at about two layers per sheet, using electron microscopy on the resulting particles. The authors have compared results with results from three additional organic solvents. They present their scalable method as potentially useful for large area applications from device and sensor fabrication to conductive composites.

A second example of direct liquid phase exfoliation to produce single layer graphene sheets is described by Mao et al. (2011). In this paper, see references therein, graphene is prepared by liquid phase exfoliation by direct sonication of graphite in the organic solvent, N-methyl-pyrrolidone (NMP). In more detail, graphite flakes with a size of 1.8–5 mm (from NGS Naturgraphit GmbH, Leinburg, Germany) were incubated in 1 mL N-methyl-pyrolidone in a 1.5 mL glass vial and sonicated for 3 h. The sonicated solution was then centrifuged at 500 rpm for 90 min. Individual graphene sheets were put onto a grid for examination in a transmission electron microscope (TEM).

It appears that exfoliation of graphite can be initiated at the edges of platelets, leaving fewer in-plane defects. A recent approach employs ball-milling of graphite with dry ice, to produce edge-carboxylated graphite (ECG) nanosheets, of size 100–500 nm, as described by Jeon et al. (2012). The resulting graphene platelets are superior to those obtained by chemical exfoliation because the interior hexagonal planes are undisturbed by the processing. The process breaks up the graphite from the edges, leaving flat hexagonal lattice planes, without the wrinkles and in-plane defects that were noted as a result of chemical exfoliation. The carboxylated edges make the ball-milled product directly dispersible in water and other polar solvents, in the form of individual planes. The carboxylated edges can later be removed by heating, e.g., after a dispersion has been deposited on the target surface. It is suggested that optically transparent films of superior conductivity can be achieved after depositing such nanosheets on surfaces, including silicon, followed by heating. It appears that this process may be inexpensive in bulk production, and that it may further the application of graphene intended as low cost electrodes. Some aspects of the ECG process are illustrated in Fig. 2.2.

The left-most image in Fig. 2.2, above, suggests that "Edge-Carboxylated Graphite" (ECG) can be reversibly obtained from graphite (center) by the ball milling procedure in dry ice. The ECG retains undefected hexagonal basal planes in platelets of size 100–500 nm, and can be reversed to pure graphite by direct annealing. In contrast (on the right) the basic chemical exfoliation route oxidizes the plane surfaces as well as the edges (a suggested structure for GO was shown in Fig. 2.1), and cannot be reversed to pure graphite by direct annealing. As mentioned earlier, when explosive exfoliation is applied to GO, reducing it to graphene, defects remain in the planes. According to Schniepp et al. (2006), to obtain single layers of GO the acid processing of graphite flakes had to last 96 h. Presumably in most cases the acid processing is of shorter duration and commonly leaves multilayers of graphite, and these multilayers are preserved in the explosive heating process. Multilayers of graphite have applications, presumably the Grafoil product is of this type. Electronic applications are more likely to require monolayer- to few-layer un-defected graphite, as is claimed to result from the ball milling process of Jeon et al. (2012).

See also, for the development of the various chemical methods, works of Stankovich et al. (2005), (2006); Dikin et al. (2007); Gomez-Navarro et al. (2007); Ruoff (2008); and Park and Ruoff (2009).

All of the above solution-based methods, chemical or liquid or edge-carboxylated exfoliations, lead to dispersions of very small 200–500 nm graphene flakes with thicknesses one to several monolayers, in units of 0.34 nm. To make a transparent conducting electrode, the tiny flakes are deposited on a substrate, typically by

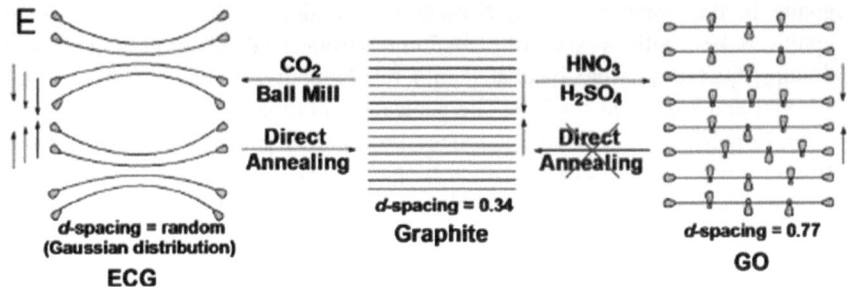

Fig. 2.2 Schematic comparison of graphite (*center*), with "edge-carboxylated graphite" (ECG) (*left*), and graphite oxide (GO) (*right*). (Reprinted figure with permission from Jeon et al., Fig. 3e. Copyright (2012) by the Proceedings of the National Academy of Sciences)

spin-coating, and the solvent driven off by heating. The resulting conductor is comprised of partially overlapping stacked graphene flakes, with little end-to-end contact, and thus exhibits significant flake-to-flake resistance. The conductivity in the normal direction that is needed for current flow in the stacked deposit is less than the c-axis conductivity in graphite, where the graphene layers are in accurate Bernal stacking, not present in the spin-coated product. The inherent weakness of electrical contact among the deposited platelets is a feature approximately independent of the in-plane defect density of the individual platelets, and stands to reduce the advantage of the less-violent exfoliation methods in the conductivity of the resulting electrode.[1]

In a recent and excellent survey of graphene electrodes by Jo et al. (2012), the solution based methods are estimated to give sheet resistances at least an order of magnitude higher than are available in electrodes made by CVD (chemical vapor deposition). In the spectrum of applications it seems clear that the lower cost of the solution-based methods will be the dominant factor in many cases. We return to this topic at the end of Chap. 3.

2.2 Direct Spontaneous Synthesis of Crystalline Graphene in Plasma and in Solution

While it has been said that graphene crystals cannot be directly grown (the restriction applies to indefinitely large crystals), spontaneous crystallizations of graphene have been demonstrated by Dato et al. (2008), Choucair et al. (2008), Lin

---

[1] There is a possibility that introducing one of the intercalant metals like K (e.g., by heating the sample in K vapor in a closed cell) during or after the spin-coating step might substantially improve the conductivity. These platelets are so refractory that simple heating does not improve their electrical contact.).

et al. (2011a), Deng et al. (2011) and Singh et al. (2011). The first report, of Dato et al., involved spraying ethanol droplets into an Ar plasma, with the results of rapid crystallization being caught on nylon membrane filters. Single layer graphene was clearly identified, along with bilayer and several-layer graphenes. The potential for large scale production is noted, with 2 mg/min of carbon being obtained for an input of 164 mg/min of ethanol.

The four following demonstrations are of spontaneous crystallizations from liquid solution, in sealed autoclave vessels, in reactions described as "solvothermal synthesis". These methods are straightforward chemistry, combining a carbon source like CCl_4 with K or Na metal to give free carbon and NaCl or KCl. The three papers mentioned use similar processes, and we describe the work of Lin et al. (2011a) as exemplary in the quality of the graphene obtained, and also in the clarity in which the method and favorable results were presented, rather closely following the earlier work of Choucair et al. The new method, that can be described as a "one pot process", is capable of bulk production of single and several layer graphene, including chemical substitutional doping of boron and nitrogen.

In what seems the clearest example, facile growth from solution of both pure- and boron-doped graphene crystals has recently been demonstrated by Lin et al. (2011a). These authors demonstrate that these graphenes form rapidly from nascent carbon and boron in reduction reactions of CCl_4 and boron tribromide (BBr_3) using alkali metal K as reductant. The products of the solution reaction appear to be free crystalline flakes of graphene and boron doped graphene, plus KCl and KBr. In a beautifully simple process to make boron-doped graphene, the authors place, into a Teflon-coated stainless steel autoclave of capacity 50 ml: 10 ml of CCl_4, 50 μL BBr_3 and 1 g of potassium K. The autoclave is sealed and heated at 160 °C for 20 h, and allowed to cool. The resultant product was dispersed in acetone under stirring to remove CCl_4 from the products. After filtering, the remaining product was washed with deionized water (1 L). The suspension was vacuum filtered and dried in a vacuum oven at 100 °C for 12 h. (The method is almost exactly that of Choucair et al. 2008, the pioneering work.) The yield of graphene was observed to be about 0.4 g C per gram of K, in both the pure and boron doped cases. This remarkable work demonstrated single layer sheets of graphene of lateral extent exceeding five micrometers, after sonication of the product described above for 40 min in ethanol. It was estimated that 13.6 % of the flakes resulting were single layer graphene. An Atomic Force Microscope (AFM) image (15 × 15 μm) in their Fig. 1d shows a 0.7 nm vertical step (indicating a single layer of graphene) at the edge of a sheet extending at least 5 μm. This process of spontaneous growth of micrometer scale single and few-layer plates of graphene in solution (without deliberate nucleation sites) is perhaps a solution analog, of the CVD process of growing carbon nanotubes using nanometer size Fe particles as catalysts. The authors use Raman spectra and measurements of the 1s XPS spectra of boron and carbon to establish that in their products the B and C atoms are in sp^2 trigonal planar bonds expected in graphene. The authors establish that the conductivity of their graphene is higher than that of reduced graphite oxide flakes that they prepared for comparison.

Singh et al. (2011) confirm a solvothermal process using ethyl alcohol as carbon source in a medium pressure autoclave. In their process at about 230 °C they note a pressure of about 60 atmospheres, leading to graphene sheets from monolayer to trilayer thickness. They use sodium borohydride as a reducing agent instead of sodium metal. They note that their process is similar to a chemical process for making carbon nanotubes as described by Wang et al. (2005).

Recently, Deng et al. (2011) grow nitrogen-doped graphene in similar reactions using an autoclave with nitrogen: for example, by reacting 20 mL of CCl_4 and 1 gram of Li_3N (a solid that melts at 813 °C) at 250 °C for 10 h. The product was washed sequentially with 18 wt % HCl, water, and ethanol, and then dried at 120 °C for 12 h. The authors state that in each batch with a 40 mL autoclave they produce about 1.2 g of carbon product. The N doping of the product was measured by XPS to be between 4.5 and 16.4 % (the latter from an augmented reaction including cyanuric chloride $(NCCl)_3$ in addition to CCl_4 and Li_3N). It appears that some of the N is not fully substitutionally bonded into the graphene, because a heating at 600 °C of the first-described product reduced its N doping from 4.5 to 3 %. In fact XPS analysis indicated two other types of N bonding, termed pyridinic and pyrrolic, in addition to the dominant graphitic bonding, in which N simply replaces C in sixfold rings. The electrical conductivity of their (heavily doped n-type) graphenes were estimated by Deng et al. (2011), using a four-point probe method, to be 0.18 Ω cm (0.15 Ω cm for the augmented preparation, with more N) compared with ~ 0.1 Ω cm obtained by Schniepp et al. 2006, for reduced graphite oxide platelets.

Returning to directly precipitated graphene, it appears that the N-doped case (Deng et al. 2011) is more complicated than the B-doped case, reported by Lin et al. (2011a), where it appears that all of the B simply replaces C in the hexagonal ring structures.

It appears that these solution-grown crystals of graphene are of higher quality than platelets obtained by any of the chemical oxidation–reduction exfoliation methods applied to graphite. It is clear that the method can be scaled to large volume. These methods seem to be promising and under-utilized.

2.3 Comparison of Chemical Graphenes to Carbon Black, Activated Carbon and Carbon Fiber

This book is concerned with applications of graphene, a material which, as we have seen in the above sections on chemical exfoliations, can appear as single to few-monolayer platelets whose lateral extent may be 100s of nanometers, as mentioned by Schniepp et al. (2006), and in the range 200–500 nm as evident in Fig. 1 of Stankovich et al. (2007). As suggested above, these lateral sizes are small compared to those available in the direct chemical growths, as exemplified by Fig. 1d of Lin et al. (2011a). Although Segal (2009) suggests that graphenes such as those obtained by Stankovich et al. (2007) and Schniepp et al. (2006) were produced in volumes 15 tons/year in 2009, with anticipated increases, these volumes are small compared to volumes of well-established commercial pure carbon products "carbon black", "activated carbon", and "carbon fiber", for which applications to some degree overlap.

These three products are based on graphitic carbon, as indicated by their relatively high electrical conductivity and black color. True amorphous carbon is a wide band gap semiconductor, insulating and non-absorbing of light. It appears that the nanocrystalline core of carbon black is on the scale of 2 nm, that of activated carbon perhaps 20 nm, while carbon fiber is made of long linear filaments bound together into long ropes or cylinders. These products vary in detail as developed for particular applications.

An exemplary form of conductive "carbon black", commercial product Cabot XC-72, was described by Muller et al. (1999) as amorphous with particle size 2 nm and surface area 250 m^2/g. Carbon black, density in the range 1.8–2.1 g/cc, is said to be produced in volume 8×10^6 tons per year, by incomplete combustion of various heavy petroleum products, called tars. Approximately 70 % of carbon black used as a reinforcing phase in automobile tires, where it facilitates heat conduction, and as a pigment. A resistivity of 0.08 Ω cm is given for XC-72 commercial carbon black, and for commercial carbon black (Alpha Aesar) a value of 1.0 Ω cm. According to Muller et al. (1999), the XC-72 carbon black of Cabot Corp. has particle size about 2 nm and surface area 250 m^2/g. A nominally similar product, Akzo Nobel EC-600JD is found by Muller et al. (1999) to have surface area 1,300 m^2/g.

As suggested above, experimental measurements show that pure amorphous carbon ("amorphous diamond"), is a wide bandgap semiconductor with tetrahedral sp^3 bonding, at density 2.9 g/cc, a density much higher than graphite but less than diamond, 3.5 g/cc. See for example McKenzie et al. (1991), their Fig. 7, a plot of the radial distribution function G(r) for undoped vacuum-arc-deposited carbon, that resembles the G(r) computed for liquid carbon. The resistivity reported by these authors, 10^7 Ω cm is much higher than for the commercial products. So it appears that commercial carbon black, some referred to as conductive blacks, is quite different from pure amorphous carbon, that is well defined and well understood in careful laboratory work with matching theoretical treatment.

A second major pure carbon product is "activated carbon", sometimes referred to as AC, that is primarily derived from coal, peat, or nut shells at a volume perhaps 2 million metric tons per year. According to Muller et al. 1999 "activated carbon" has a higher surface area, in the range 500–1,500 m^2/g (compared to completely exfoliated graphite, 2,630 m^2/g) and has widespread use for sorbing undesired chemicals as in water purification, air purification and as an antidote for swallowed poison.

A careful TEM study (Harris et al. 2008) of an exemplary activated carbon, Norit GSX powder, derived from peat (a recent biological residue) found that the particles of porous powder were disordered, but composed mostly of tightly curled single carbon layers with pore size about 3 nm. Harris et al. (2008) describe Norit GSX, also described as activated charcoal, as produced from peat, carbonized in inert atmosphere at 500 °C, followed by activation by heating in steam at 1,000 °C, and washing in HCl. Norit GSX has a surface area, according to Harris et al. (2008), of 950 m^2/g. According to Gamby et al. (2001), Norit activated carbon (AC) powder has a surface area 1,200 m^2/g and resistivity 3 Ω cm. (It is tempting to attribute this higher resistivity, compared to the $\sim$0.1 Ω cm of the reduced graphite oxide reported by Schniepp et al. 2006, to the smaller lateral extent of the graphitic planes, $\sim$20 nm in activated carbon to the 200–500 nm of lateral extent in the exfoliated products.) Gamby et al. (2001) found that a similar powder, PICACTIF SC carbon, had a specific capacitance of 125 F/g, which is of interest for supercapacitor application (to be discussed in Chap. 3).

Activated carbon is obtained from sources that are described as "non-graphitizing" (Franklin 1951), meaning that they cannot be turned entirely into six-fold bonded carbon even by heating to 3,000 °C. Harris et al. take the position, perhaps somewhat controversial, that the non-graphitizing form of activated carbon includes 5-fold rings (as do the curved fullerenes) that, once formed, cannot be removed. Harris et al. suggest that the 5-fold rings account for the curvature of the single carbon sheets seen in their TEM image of Norit GSX, and correlate with the high surface area.

"Carbon fiber", the additive for a major contemporary structural material, the "carbon fiber composite", is composed of larger strands of graphitized carbon 6–10 μm in diameter, produced either from petroleum pitch or from polyacrylonitrile (rayon), where the stacking of parallel layers is turbostratic (random) rather than of the Bernal AB type in the petroleum pitch products. Carbon fiber-reinforced polymers are important in light-weight structural materials, used, for example in helicopters, jet aircraft and bicycle frames, where light weight, strength and flexibility are important. An estimate of the market for carbon fiber composites in 2009 is $11 B. Use of carbon fiber in expensive automobiles is described by Brown (2013), who points out that the cost in this application is mostly in the processing, explained in some detail, rather than in the raw material. Only a small part of the carbon fiber market seems susceptible to graphene as an alternative, but Head, manufacturer of expensive tennis racquets as well as of expensive skis, has announced new racquets that incorporate, in their words, "the world's strongest and lightest material".

These small particle forms of carbon, apart from activated carbon, are used as fillers in a range of applications. Carbon black is used in tires, in nanocomposites to impart strength and electrical conductivity, and may be incorporated in the skin of stealth aircraft to absorb radar pulses. Compared to carbon black, it seems that the larger platelets of graphene will be more expensive to produce and may likely only find reasonable application where the higher electrical conductivity is important. The traditional forms of small particle carbon, carbon black, activated carbon and carbon fiber, with large volumes and low commodity prices, are not suitable for transparent electrodes, nor for electronic logic, where the need for plate geometries with high aspect ratio clearly leaves open new applications for graphene. It remains to be seen whether the small platelets of graphene will form adequate electrodes for solar cell and other electrode applications. An important issue is the electrical connectivity between platelets, that may not suffice for some electrode applications, although it may suffice for electromagnetic shielding applications.

Higher level electronic applications can be addressed by large area continuous polycrystalline graphene, that can now be made by chemical vapor deposition, CVD.

2.4 Chemical Vapor Deposition-Based Methods CVD

It has been known for some time that graphene growth from carbon-containing gases can be catalyzed by various transition metal substrates. This can be described as catalytic cracking of a hydrocarbon source such as methane or acetylene, or precipitation of dissolved carbon (perhaps from a cracking process) onto a metal surface with subsequent graphitization (ordering into the hexagonal lattice). Reports include Grant and Haas (1970); Gall et al. (1987); Nagashima et al. (1993); Gall et al. (1997); Forbeaux et al. (1998); Affoune et al. (2001); Harigaya and Enoki (2002). Limited solubility of C in the substrate can restrict the production to single layers of graphene, a desirable outcome.

An important discovery of Reina et al. (2008) was that single crystal (and, later, also continuous polycrystalline CVD) graphene could be released from the traditional oxidized silicon substrate to be relocated to a different target substrate.

The steps used by Reina et al. (2008) include covering (spin-coating) the film with poly(methyl methacrylate) (PMMA, 950,000 molecular weight, 9–6 wt % in anisole), followed by partially etching the surface of the SiO_2 in 1 M NaOH aqueous solution. The 300 nm SiO_2 does not etch completely, and only a minor etching is enough to release the PMMA/graphene layer from the silicon. The release is typically accomplished by placing the substrate in water at room temperature, where manual peeling can be used to detach the PMMA/graphene membrane from the substrate. As a result, a PMMA membrane is released with all the graphite/graphene sheets attached to it. This membrane is laid over the target substrate, and the PMMA is dissolved carefully with slow acetone flow. The flakes in the original work of Reina et al. (2008) are on the micrometer scale, obtained from micromechanically cleaved graphite.

This basic transfer technique was quickly adapted to much larger area graphene films grown by chemical vapor deposition CVD on transition metals, principally Ni and Cu.

Li et al. (2009) extended this basic method of growing graphene on copper foil, followed by transfer using the Reina method, to the centimeter scale, transferred to a Si substrate. Li et al. (2009) use chemical vapor deposition with methane as the carbon source, flowing over copper foil about 25 μm in thickness at temperatures up to 1,000 °C. It was found that the growth was complete in 30 min, and was self-limiting, so that only 5 % of the area had more than one layer of graphene. The graphene was found to be continuous, growing across lattice defects in the copper substrate. The mobility in the grown films was measured as high as 0.4 m^2/Vs. The films were transferred to arbitrary substrates by a method similar to that of Reina et al. (2008).

Reina et al. (2009) also extended their method (Reina et al. 2008) to growing graphene on Ni, but it now appears that Cu is more suitable for this purpose. Reina et al. (2009) include references to earlier reports of growth of graphene mono-layers on single crystalline transition metals, including Co, Pt, Ir, as well as Ru, Ni, and Cu.

Large 4 × 4 in.^{2} graphene films grown on copper and transferred to oxidized silicon were reported by Cao et al. (2010). They found carrier mobility about 0.3 m^2/Vs and observed both the quantum Hall effect and weak localization in measurements of the graphene.

A similar method based on Ni substrates has been described by Kim et al. (2009). These workers have produced patterned graphene by patterning a Ni deposit on a Silicon wafer ahead of the CVD growth of graphene. They also described several methods of transferring the Ni grown graphene to other sub-strates. Growth on Co has recently been presented by Ramon et al. (2011).

Plasma-enhanced CVD (PECVD) has been applied to growth of carbon nanostructures by several authors (Zhao et al. 2006; Wang et al. 2004; Kobayashi et al. 2007 and Chuang et al. 2007). It is not clear if the plasma-assisted aspect could be applied in the wide area growth of graphene on Cu or Ni, to allow a lower temperature deposition that clearly would be desirable.

The state of the art in growing and transferring large area graphene films is reported by Bae et al. (2010), a consortium of workers in South Korea. Bae et al. have achieved deposition on a roll of Cu foil with dimension up to 30 in. in the diagonal direction. The Cu foil is wrapped onto a cylindrical quartz substrate in the CVD reactor, so that a larger graphene area can be obtained within a tube reactor. The copper foil was wrapped around a 7.5 in. diameter quartz tube, inserted into an 8-inch-diameter quartz reactor tube of length 39 in. Adopting this geometry minimized temperature gradients over the copper area, to promote homogeneity of the deposited carbon. The authors describe a preliminary step of annealing the copper foil in flowing H$_2$ at 1,000 °C to increase the grain size of the Cu from a few μm to around 100 μm. After this annealing, methane is added to the gas flow for 30 min at 460 mTorr at flow rates 24 and 8 standard cc per minute, respec-tively, for methane and hydrogen. After this the furnace is quickly cooled to room temperature at 10 °C/s in flowing hydrogen at 90 mTorr.

The graphene-coated large area copper foil, after removal from the quartz tube, is attached to a "thermal release tape", a large-area thin polymer supporting sheet, by running the pair of sheets between soft rollers at a pressure ∼2 MPa. The result, a large-area, "release tape" polymer/graphene/copper sheet sandwich is then rolled through a chemical bath to remove the copper. The next step is to place the graphene side of the assembled sheet onto the substrate of final choice, that is often a roll of 188 μm thick polyethylene terephthalate (PET). Running this sandwich between rollers at mild heat ~ 100 °C transfers the graphene from the "thermal release tape" to the desired final support. In typical cases the procedure is iterated to give a stack of four graphene layers (whose azimuthal orientations are random) on the wide area PET supporting sheet, and with diagonal dimension as large as 30 in. This stack is a high quality transparent flexible conductor at best with ~ 90 % optical transmission and low resistance of ~ 30 Ω/square. This low value was aided by a step of "chemical doping" with nitric acid HNO_3, that makes the films strongly p-type. According to the authors, Bae et al. (2010), this transparent conductor is superior to common transparent conductors such as Indium tin oxide (ITO) and carbon-nanotube films, reported by Lee et al. (2008). Bae et al. (2010) describe steps in producing touch screens for electronic devices using their graphene/PET transparent electrodes. The resulting transparent conductors display extraordinary flexibility, and it was stated that the electrode functions up to 6 % strain.

Aspects of the growth of graphene on Cu that may be improved are the high temperature, 1,000 °C, conventionally needed, the rather small size of the polycrystalline graphene grains typically resulting, and, of course, the problem of removing the copper from the graphene.

To begin, note that the interaction of carbon with copper is much weaker than with carbon (differing from Ru, where epitaxy of C is achieved, with corrugations of the graphene to match the lattice). The Cu does nucleate islands of graphene, but without strong relations between the axis directions in the graphene and the copper.

The growth of graphene on Cu foil, typically of (100) surface orientation, is successful, but leads to lower mobility than that of micromechanically cleaved graphene. Intuitively it seems that grain boundaries are likely the cause of lower mobility, although it may be affected by lack of cleaning of adsorbates or stray electric fields from the support of the graphene during the measurements. A theoretical discussion of growth on Cu (111) has been given by Chen et al. (2012) that seems useful even though the copper foils in use have (100) surface orientations.

Li et al. (2011a) report growing "single crystals" of graphene on Cu (100), of 0.5 mm size, remarkably large indeed, and larger than grains in the underlying Cu. They comment that the employed foil displayed "a highly faceted, rough Cu (100) surface with sharp diffraction spots". They also comment that the diffraction pattern of the graphene looked like that of free standing graphene, "perhaps indicating a weak coupling to the rough surface". In fact the mobility in their "single crystals" was 0.4 m^2/Vs, very high, but not a single crystal value as obtained on free cleaned graphene. A method of growing oriented single crystal graphene on Cu without grain boundaries would be an important advance. Chen et al. (2012), presuming Cu (111), note that this surface has hexagonal symmetry matching the graphene benzene ring symmetry. (Cu 111 may be easily available, as it results from evaporation of Cu, as shown by Tao et al. (2012), allowing in CVD growth also excellent monolayer graphene.) On Cu (111), however, Chen et al. find in simulation that a typical nucleated seven-ring cluster (a ring surrounded by six rings, this resembles the coronene molecule), initially benefitting

from the common hexagonal shape, in fact reduces its energy by rotating 11°
relative to the hexagon of the underlying Cu. For this reason Chen et al. (2012)
argue that orientational disorders of carbon islands will be abundant in the early
stages of nucleation and growth on Cu (111). Such disorders cannot heal with the
enlargement of the islands, leading to the prevalence of graphene grain boundaries
upon island coalescence. Chen et al. go on to suggest that by adding Mn to the Cu,
a Mn-Cu (111) surface of hexagonal symmetry could easily be obtained that would
exactly stabilize the common "coronene" seven-ring precursor. They in fact
suggest that molecular coronene vapor be used in a two-stage CVD process (it has
been shown by Li et al. (2011b) that benzene, when used in the CVD process on
Cu, allows graphene growth at 300 °C, much lower than the typical 1,000 °C). The
suggested coronene deposition, followed by de-hydrogenation, would leave a
network of like-oriented graphene nuclei that would then be filled in, in a second
stage, by CVD of methane or ethylene in the usual way. In this two stage growth,
on a modified Cu (111) surface, it appears that grain boundaries would not arise
and a graphene single crystal could be envisioned of very large lateral size and
controlled orientation, as suggested by Chen et al. (2012). As noted above, Li et al.
(2011a) have already reported growing single crystals as large as 0.5 mm, on Cu,
but without control of the orientation of the graphene.

These reports are promising regarding the chance of larger grain size and also
lower deposition temperature for graphene grown on copper.

The problem remains that Cu is not a useful substrate in most cases. It would be
helpful to find ways to grow graphene on silicon, for example. A preliminary
report of UHV growth of graphitic carbon directly onto Si (111) by e-beam
deposition was given by Hackley et al. (2009). The silicon surface temperature was
initially 560 °C, leading to a thin amorphous carbon layer, and then raised to
830 °C where graphitic carbon was observed to grow. The grain size of graphitic
carbon was in the one nanometer range in this report, which is not very promising.

An interesting chemical approach to growing reproducible graphene nanoribbons has been described by Koch et al. (2012). These workers fabricated long
narrow armchair graphene nanoribbons on a gold surface, Au (111). The processing was done in ultrahigh vacuum (UHV), evaporating the 10,10'-dibromo-
9,9'-bianthryl (DBDA) precursor molecules from a Knudsen cell at 470 K. The
word "bianthryl" indicates a molecule composed of two anthracene molecules,
and anthracene in turn is three benzene rings in a row. These molecules polymerize
in a linear fashion on the Au surface, to lead, finally, to chemically-synthesized
nanoribbons with perfect armchair edges. Thus, the ribbons alternate between two-
and three- benzene rings in width. Koch et al. used STM to study the "voltage-
dependent conductance" as single ribbons are pulled free of the gold surface.
While the authors do not describe the nanoribbons as metallic, they did observe
currents up to 100 nA through these structures. The large current density suggests
that the nanoribbons support metallic conduction in a 1D band above a band-edge
energy. The ribbons were mechanically robust and were moved around the gold
surface using the STM tip. The methods are attractive in achieving long ribbons of
precise width and edge structure, and also in that the highest temperature in the

fabrication is only 400 °C. Producing wider ribbons and also finding ways to locate the ribbons where they are needed are issues facing application of this method.

2.5 Silicon Carbide Epitaxial Growth

Graphene with long range order can be obtained by heating SiC single crystals in vacuum. Wafers of SiC are commercially available, and form the basis for graphene transistor fabrication. Silicon carbide can be viewed as a layer compound, composed of polar SiC layers. The bonding in SiC is principally covalent, as in diamond C and in Si, but the elemental difference imparts some ionic character. The polar tetrahedrally-bonded layers of importance are arranged perpendicular to the c(0001) direction. There are two types of stacking of the SiC layers, cubic stacking and hexagonal stacking. The important hexagonal "polytypes" 4H and 6H, comprise, respectively, 4 and 6 such hexagonally stacked bilayers per unit cell. In 4H-SiC and in 6H-SiC the Silicon polar surface is designated "0001" while the polar Carbon surface is designated $[000(-1)]$. All of these crystals are insulators, with bandgaps 3.02 eV for 6H-SiC and 3.27 eV for 4H-SiC. (Seyller 2012). Graphene layers can be epitaxially formed by heating SiC in vacuum, as Si atoms leave and the surface reconstructs. A detailed procedure has evolved using the silicon face SiC(0001).

Growth of graphene by heating these surfaces was first demonstrated by Van Bommel et al. (1975), and confirmed by Forbeaux et al. (1998), who show 6-fold LEED pattern of crystalline graphite after annealing 6H-SiC(0001) surface at 1,400 °C, observed with primary energy 130 eV, The growth methods were recently reviewed comprehensively by Seyller (2012). An earlier review of growing graphene and bilayer graphene on SiC was given by Bostwick et al. (2009), who also make extensive use of the surface technique angle resolved photoemission spectroscopy (ARPES).

The electronic band structure of graphene grown on SiC has been investigated and found to be nearly ideal by Sprinkle et al. (2009). Their methods include ARPES and surface X-ray diffraction (SXRD). Studies of graphene grown on SiC have revealed that even multi-layer growths retain single layer electronic properties (Hass et al. 2008). It is generally found that graphene on SiC is conductive with electrons, an n-type film with Fermi energy in the order of 0.3 eV.

While the SiC crystalline substrate imparts long range order to the grown graphene, the resistivity of the grown layers is not as low as obtained in flakes micro-mechanically exfoliated from graphite. A leading cause of this degradation appears to be steps on the SiC face, whose effect on the resistance of epitaxially grown single layer graphene was investigated by Ji et al. (2012). The effect of the steps is apparently not to disrupt the order of the grown graphene but to cause jumps in the doping level.

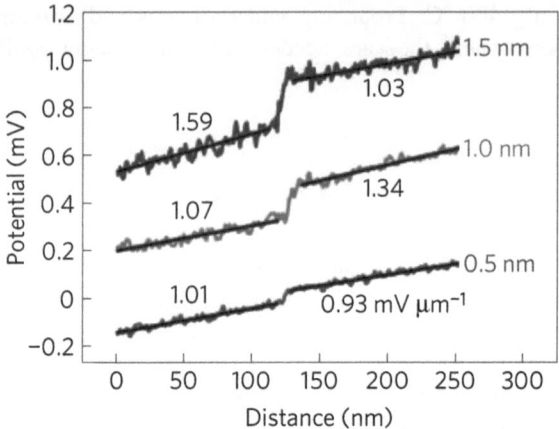

Fig. 2.3 Scanning tunneling potentiometry measurements on single layer graphene on SiC in traces (*top* to *bottom*) crossing steps of heights 1.5, 1.0 and 0.5 nm, respectively. The numbers shown close to the data traces indicate the slopes in mV/μm on the terrace before and after the step is encountered. The authors confirmed that the voltage drops across the steps scale with the injected current. (Reprinted figure with permission from Ji et al., Fig. 3c. Copyright (2012) by Nature Materials)

The Scanning Tunneling Potentiometer device of Ji et al. (based on that of Bonnani et al. (2008), see also Homoth et al. (2009)) deploys three tips: two are current-injecting probes in contact with the surface, typically separated by ∼500 μm, and the third is a scanning tip. Data from this method is given in Fig. 2.3.

The scanning tip measures the topography as well as the local potential, as affected by the imposed current flow. The parameter values relevant to the Fig. 2.3 are temperature 72 K, voltage drop between current injecting tips 1.53 V, the estimated local current density in the single monolayer graphene layer in the measured region, 6.4×10^{-6} A/μm. The carrier density in the graphene is estimated as 10^{13} cm^{-2}, and the mobility is estimated as 0.3 m^2/Vs.

The authors find that voltage-drops across steps contribute significantly to the total drop across a sample. A 0.5 nm substrate step contributes extra resistance equivalent to a terrace about ∼40 nm wide, while 1.0 and 1.5 nm high steps contribute resistances, respectively, equivalent to terraces ∼80 nm wide and ∼120 nm wide.

Realistic calculations of the extra electrical resistance that appears when graphene "flows" over a step of height h_s on a SiC substrate, have been carried out by Low et al. (2012), as suggested in Fig. 2.4. (This situation occurs in device applications as reported by Lin et al. (2010), to be discussed in Chap. 5, see Fig. 5.1. More work on related step-geometries has been carried out by Sprinkle et al. (2010) and by Hicks et al. (2012), to be described in Chap. 4.) Referring to Fig. 2.4, the "step-resistance effect", as interpreted by Low et al. (2012), will vary according to the degree to which the substrate in question changes the carrier

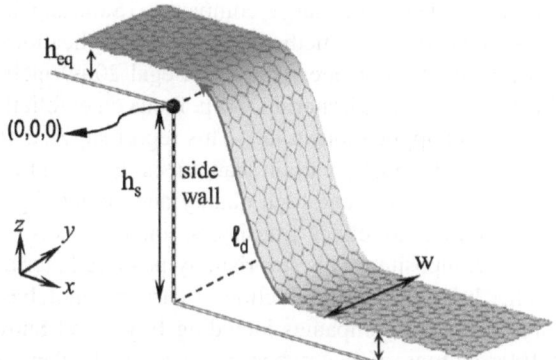

Fig. 2.4 A large incremental resistance is measured when graphene "flows" over a step-edge on SiC, as illustrated here. The effect comes from the large difference in local Fermi level in the bound and free sections of the graphene and, according to Low et al. (2012), is not influenced strongly by the curvature. (Reprinted figure with permission from Low et al., Fig. 1. Copyright (2012) by the American Physical Society)

density of the graphene, as well as the density of steps on the substrate. (Hicks et al. (2012) have patterned deeper trenches into SiC, where they believe the graphene (subsequently grown by sublimation heating) on the trench wall, analogous to the "flowing" graphene section in Fig. 2.4, actually develops a bandgap on the order of 0.5 eV. These authors find that a locally observed energy gap is confined to regions of sharp curvature in the graphene.). The equilibrium graphene spacing from the SiC substrate is taken as 0.34 nm, with an estimate of 0.04 eV per atom binding by van der Waals interaction (Zacharia, et al. 2004).

To summarize, a lower graphene electron mobility for epitaxial films on SiC compared to mechanically exfoliated graphene placed on oxidized Si, has been connected to the presence of steps on the SiC surface. But the calculations carefully performed by Low et al. suggest that the sharp curvature suggested in Fig. 2.4, as the film flows over the step-edge, is not the source of much scattering. Rather, the extra scattering, is due to the electrical coupling between the graphene and the substrate, that varies sharply in the vicinity of the step. The authors conclude that the morphology affects the resistivity through the tendency of the SiC to dope the graphene, an effect stronger on SiC than on oxidized Si.

Adapting the SiC graphene process toward a production method, Emtsev et al. (2009) have shown that it can be carried out at atmospheric pressure, producing wafer-size layers with good results. De Heer et al. (2010) (and references therein) have demonstrated an improved form of epitaxial graphene growth: "confinement-controlled-sublimation", including methods to produce step-free surfaces on SiC.

A useful summary of synthesis methods for graphene with emphasis on patent applications filed has been given by Samba Sivudu and Mahajan (2012). A summary of this report was given on the nanowerk website dated June 28, 2012, http://www.nanowerk.com/spotlight/spotid=25744.php

The patent survey suggests that large companies (Samsung leading with 16 patent applications) focus on CVD methods (92 patent applications total) suitable for electronics, while small companies (see also Segal 2009) including Angstron Materials, XG Sciences, Vorbeck Materials Corp, focus on exfoliation methods for small platelets (94 patent applications total). This report suggests that nanoribbon manufacture based on unzipping carbon nanotubes is a synthesis method of "moderate scalability, high yield and high quality and potentially low cost" with potential applications in field effect transistors, interconnects, nanoelectromechanical devices and composites. This surprisingly positive latter assessment may build from the rather large present production of carbon nanotubes. According to Segal (2009), large chemical companies including Bayer and Showa Denko produced in 2009 100s of tons of nanotubes per year, with plans at that time to increase production to thousands of tons.

Chapter 3
Solar Cells and Electrodes

An important potential application of graphene is as a component of a solar cell. Highly conductive, transparent graphene can serve as one or both electrodes, one of which has to let light into the absorbing region of the device. The photovoltaic action of a solar cell occurs as photo-generated carriers, electrons and holes, are generated in (or flow into) a central region of strong electric field, that sends carriers of opposite charge in opposite directions. In the conventional silicon solar cell, the absorbing regions extend beyond the depletion region, containing the electric field, into the oppositely doped electrode regions, by diffusion lengths L on either side.

3.1 Solar Cell Concepts

Graphene cannot play an important role in absorbing light, but graphene electrode layers in a graphene-absorber-graphene cell can be doped oppositely, that can serve both to increase the conductivity and to generate an electric field across the absorber. Sketches of the semiconductor and organic types of solar cell are shown in Fig. 3.1.

Favorable aspects of graphene for solar cell application are its transparency and high conductivity; its durability, and the facts that carbon is neither rare nor toxic. At the moment there appears to be no commercial solar cell product using graphene. This situation could change if a market were identified strongly favoring a fully flexible solar cell, or if more conductive chemically-derived graphene layers or less expensive epitaxially-produced graphene layers were to become available. There are large scale projections for providing non-fossil-fuel energy for large populations, and these projections depend heavily on solar power and wind power in a sufficiently large grid that the problem of storage of energy is minimized (Delucchi and Jacobson 2011). The recent detailed proposal for entirely powering the State of New York by Wind, Water and Sun (WWS) by the year 2030 (Jacobson et al. 2013) assumes 4.97 million 5 kW residential rooftop solar installations (assuming Sun Power E20 20 % efficient silicon solar cells), but notes that the cost of solar power with these cells is high, relative to the wind-power portion of the plan. The New York

E. L. Wolf, *Applications of Graphene*, SpringerBriefs in Materials,
DOI: 10.1007/978-3-319-03946-6_3, © Edward L. Wolf 2014

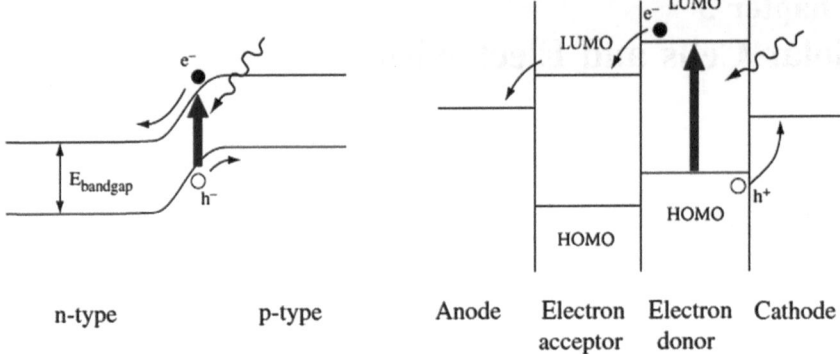

n-type   p-type   Anode   Electron Electron Cathode
                                                 acceptor   donor

Fig. 3.1 *Left* Band diagram of semiconductor pn junction solar cell. Depletion region provides electric field to separate carriers electrons and holes, that can in fact be absorbed in a wider region extended by the diffusion lengths for minority holes and minority electrons. *Right* Organic solar cell comprises Electron donor molecule that absorbs light, creating a hole in the HOMO (highest occupied molecular orbital), that transfers to the cathode. The photo-electron transfers to the lowest unoccupied molecular orbital LUMO of the electron acceptor, and thence to the anode. Graphene can be used for anode and/or cathode in such a device, and variations on the light-absorbing portion of the cell include dye sensitization and insertion of quantum dot absorbers. (Reprinted figure with permission from Bosshard, Fig. 7. Copyright (2006) by the Global Climate and Energy Project, Stanford University)

State plan also calls, by 2030, for 49.7 GW (49.7×10^9 W) of installed solar capacity in commercial or government roof PV systems, and 41.4 GW installed capacity in solar PV plants. The total solar capacity called for in this comprehensive plan, for the State of New York, is thus 116 GW by 2030. This plan may be unlikely to be implemented in the near term, but does suggest the very large market that may appear at least in some parts of the world.

Large solar-cell projects are underway in other locations, sometimes to fill in power at midday peak usage and sometimes in conjunction with wind farms. A plan of combined wind and solar power is being implemented by the Chinese government in Mongolia, a region of high sunlight and high wind velocity. The contractor for the photovoltaic portion of this project is First Solar, the largest thin film solar cell supplier, manufacturing thin film cells of CdTe. According to the New York Times (Woody 2009) First Solar, an American firm based in Tempe Arizona, signed an agreement with the Chinese government for a 2 GW photo-voltaic farm to be built in the Mongolian desert. The photovoltaic farm, of area 25 square miles, is part of a 11.9 GW renewable energy park to be built at Ordos City in Inner Mongolia. The overall project is to include 6.95 GW of windpower, 3.9 GW of photovoltaic power, and 0.72 GW of solar thermal farms. First Solar is likely to build a plant in China to make thin-film solar panels. According to the article, the 2 GW solar farm as built in China is likely to be significantly less than the $5–$6 B if it were built in the US. It is commented that the CdTe solar cells of First Solar are less efficient than the standard crystalline silicon solar cells such as the E20 cells of Sun Power, but they are significantly less expensive to build.

First Solar has also recently agreed to provide 1.1 GW of capacity to two California utilities from three big solar farms. It is commented in this article that the Chinese project is atypical of large solar projects, which have generally been awarded to solar thermal technology, which deploys mirrors to heat a liquid to create steam that drives an electricity-generating turbine, rather than to straight photovoltaic projects. It is pointed out in this comparison that the straight photovoltaic projects generally have fewer environmental impacts, such as requiring cooling water, and can be brought online faster than solar thermal plants.

The solar cell market is extremely large and likely to grow. The chance for graphene-containing cells to compete in this market, perhaps small, seems worth consideration, in view of the potential market size.

3.2 Organic Solar Cells

Organic photovoltaic cells, similar to the right panel in Fig. 3.1, based on solution-derived graphene deposited on quartz, were described by Wu et al. (2008). In these solar cells the layer sequence is graphene, copper phthalocyanine (CuPc donor)/fullerene (C_{60} acceptor)/bathocuproine (BCP), Ag (1,000A). (In comparison cells the quartz-graphene layer was substituted by a conventional transparent conducting electrode, 130 nm thick ITO (Indium tin oxide) on glass.) To prepare the quartz-graphene transparent anode, graphite oxide flakes, obtained by the Hummers acid oxidation of graphite, were spun from dispersion onto the quartz. Heating in vacuum was essential to restore conductivity to the graphene layer. This chemical preparation was similar to that later used by Wu et al. (2010), but the stated resistance per square, for <20 nm thick graphene, was quoted in the range 5 kΩ – 1 MΩ, much higher than the 800 Ω/square given by Wu et al. (2010). The conversion efficiency of these cells (0.4 %) was rather poor, inferior to the comparison cells on ITO (0.84 %). This deficiency was partly attributed to the high sheet-resistance of the graphene, a feature improved upon in later work by the same authors. The early report of Wu et al. (2008) may be more of an indication of learning curves in processing chemically-derived graphene and in its incorporation in organic solar cells, than that graphene films, even those chemically derived, cannot be envisioned in this application. For example, doping the graphene layer to increase its conductivity was not attempted.

3.3 Cells Utilizing Dye Absorbers

Dye molecules are incorporated into wide-bandgap semiconductor photovoltaic cells to extend the absorption into the red portion of the solar spectrum. Dye-sensitized solar cells using graphene as a transparent electrode have been reported by Wang et al. (2008a) and by Eda et al. (2008).

As background on dye-sensitized cells, the combination of TiO_2, tetragonal anatase, with a wide bandgap, and dye has been used as absorber in a class of inexpensive solar cells, reviewed by Graetzel (2001). In this approach the absorber (the central portion of the right panel in Fig. 2.3) is partly titanium oxide, TiO_2 in its anatase tetragonal form, and partly dye molecules. Anatase can be prepared in a nano-porous form (sometimes described as mesoporous, although microscopy shows particles of 10–80 nm diameter) by various procedures, one described as hydrothermal processing of a TiO_2 colloid.

This oxide has a bandgap 3.2 eV, which means that the maximum light wavelength absorbed is 388 nm = (1,240 eV-nm)/(3.2 eV), that is much shorter than the peak wavelength in the solar spectrum, about 484 nm. Thus only a small part at the UV end of the solar spectrum is absorbed. Spectacular extension of the light absorption range to at least 700 nm has been demonstrated by coating (sensitizing) the nanoporous anatase with dye. The dyes have a relatively low specific absorption, so that a thick layer of dye is needed. Each dye molecule should be in intimate contact with the anatase surface so that a photoelectron can reliably be transferred to the conducting anatase. The traditional transparent electrode in dye-sensitized cells has been indium tin oxide ITO (or related FTO fluorine tin oxide), on which the anatase layer is deposited, followed by the dye.

Graphene transparent electrodes (chemically exfoliated) were applied to dye-sensitized solar cells by Wang et al. (2008) and by Eda et al. (2008). Sheet resistances around 1 kΩ/square were obtained by Wang et al. (2008) after spin-coating graphite oxide particles and reducing the deposit by heating to 1,100 °C. The band diagram of the cells proposed by Wang et al. (2008), is indicated in Fig. 3.2.

The authors describe the TiO_2 layer (See Fig. 3.2) as a "hole-blocking" layer, so that holes created in the dye can only flow to the Au cathode. The operation of the device is described as a flow of electrons, photo-generated in the dye, into the conduction band of the TiO_2, followed by their flow to the graphene electrode by a percolation mechanism in the mesoporous TiO_2 structure. The photo-holes created in the dye flow to the Au cathode, a process also described as regeneration of the photo-oxidized dye, by electrons from the hole-conducting spiro-OMeTAD molecules. The open-circuit voltage of the device at full illumination was measured as 0.7 V. In open circuit under illumination, the anode electron energy levels rise to counter the flow of electrons from the excited state of the dye, and the cathode Fermi level tends to align with the ground-state electron level in the dye. The maximum open circuit voltage is then approximately 1/e times the energy level difference in the dye molecule. The absorption of the dye molecule peaks near 600 nm, (where the anatase has no absorption) that would suggest the energy level difference in the dye as $\Delta E = hc/\lambda = 1,240$ eVnm/600 nm = 2.07 eV. This is nearly three times the observed open circuit voltage, suggesting that there are additional voltage drops within the structure under illumination. (There must be energy drops as indicated in Fig. 3.2, between the dye molecule's excited state and the anatase conduction band, and between the anatase conduction band and the graphene, to produce a short-circuit current under illumination, as observed.)

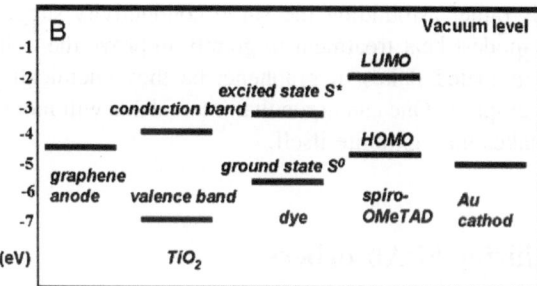

Fig. 3.2 Band diagram of graphene-anode dye-sensitized solar cell of efficiency 0.26 %, using a ruthenium-based dye. In this diagram "Spiro-OMeTAD" is a hole-conducting molecule. (Reprinted figure with permission from Wang et al., Fig. 3b. Copyright (2008) by the American Chemical Society)

Nonetheless, 0.7 V is a respectable open-circuit voltage. Unfortunately, the corresponding conversion efficiency is low, reported as 0.26 %, with a higher value 0.84 % obtained in a similar structure using the conventional transparent electrode FTO fluorine tin oxide in place of the chemically produced graphene. The authors suggest that difficulty lies in the series resistance of the device, originating in interfacial voltage drops and the lower conductivity of the graphene anode. The performance at present is poor, for example, as compared to the Sun Power E20 panels available in quantity at 20 % efficiency.

3.4 Cells Utilizing CdTe Absorbers

A direct chemical method for making boron-doped graphene flakes (Lin et al. 2011a) was described above. These authors also show that their flakes can be used to make improved back-contacts to CdTe solar cells. This application was not to provide a transparent electrode, but simply to use the strongly p-doped graphene, applied as a paste, to suitably contact the p-type CdTe in these cells. The transparent front contact in their cells remained as FTO, fluorine-doped tin oxide on glass. It appears that wider uses to the directly synthesized graphenes in solar cell design may soon appear, that would also use the transparency of graphene as an electrode. Also, in connection with the electrical conductivity of the spin-deposited-flake electrodes of graphene, the inter-flake electrical coupling that is a limiting factor might possibly be improved by heating the deposit at modest temperature in potassium vapor. For example, Pureval (2010), citing earlier work, indicates that the compound KC_8 forms in a closed two-zone cell with graphite held at 300 °C and potassium held at 250 °C. In a related much earlier work, Vogel (1977) reports that graphite intercalated with SbF_6 exhibits extremely high (p-type) conductivity $10^6 \ (\Omega \ cm)^{-1}$, about 40 times the conductivity of pure graphite and 1.5 times the conductivity of copper metal! (SbF_6 is a liquid that boils at 150 °C.) This rather astounding measured conductivity suggests flexibility in searching for a modest heat treatment to greatly improve the c-direction conductivity of the spin-coated flakes of graphene, be they chemically synthesized or exfoliated from graphite. One can argue that intercalants will more easily permeate the deposit of flakes than graphite itself.

3.5 Cells Utilizing Si Absorbers

The market-dominating silicon solar cell is a pn junction with a thin highly-doped n-layer, the front, light-admitting electrode, on a p-type substrate. Light entering at the n-layer is partially absorbed in the diffusion layer adjacent to the depletion region that separates the photocharges, as well as in the p-layer behind the junction. Drawbacks of the silicon solar cell are the relatively weak absorption (because of the indirect band-gap) requiring a thicker layer, and the necessary high-temperature diffusion of donor impurities to form the pn junction. A simplification of such a junction is the Schottky barrier cell, in which a thin semitransparent metal is deposited directly on the (n-type) silicon surface, leading to a positively-charged depletion region as donor electrons fall into inherent empty surface states at the silicon/metal interface. A variation of a Schottky barrier solar cell is one in which the metal layer is replaced by graphene, as recently reported by Miao et al. (2012). In that work, n-type silicon was directly covered with graphene. Superior solar cell performance was achieved in devices where the graphene, grown by CVD on copper foils and transferred to the n-silicon surface, was subsequently doped with TFSA that makes it p-type, lowering the Fermi level into the valence band. Thus, TFSA, bis (trifluoromethanesulfonyl-amide [$((CF_3SO_2)_2NH)$]) transfers positive charge to the single graphene layer, increasing its work function. The clean surface of Si exhibits a large density of localized surface states near midgap, to which electrons from donor atoms in the Si transfer, leaving a positive charge density and Schottky barrier in the Si. This effect is, to some extent, modified by the nature of the metal (or graphene) placed onto the clean silicon surface. The resulting built-in potential of the barrier, defining the maximum available open-circuit photovoltage, separates the holes and electrons in the functioning of the solar cell. Quite ideal solar cell I(V) curves are shown in the paper of Miao et al. (2012). The efficiency of their best cells is 8.6 %, considerably better than in the absence of doping of the graphene. The authors suggest that the same direct transfer of graphene, to make a Schottky barrier, with results enhanced by TFSA doping, might be applicable to make Schottky barrier solar cells on other semiconductors, including CdS and CdSe. Such cells might be less expensive than silicon. The paper gives experimental details, and also, references to earlier work. The use of CVD-grown graphene, rather than any type of chemically-prepared graphene, will add to the cost of such cells. It is also possible that useful results could be obtained using less-expensive graphene from chemical exfoliations or from direct synthesis, as described in Sect. 2.2.

Optimizing graphene electrodes by varying the charge concentration and moving the Fermi energy is important in solar cell applications of graphene. Doping by organic polymers, although not in a solar cell context, has been recently reported by Sojoudi et al. (2012), who also confirm the importance in practice of p-type doping by typical atmospheric exposure (Schedin et al. 2007). Sojoudi et al. use amine-rich APTES [3-Aminopropyltriethoxysilane] for electron doping and fluorine-containing Perfluorooctyltriethoxysilane (PFES) for hole doping. They actually transferred CVD-grown graphene from a copper surface onto a Si/SiO_2 substrate, further prepared with adjoining self-assembled monolayers (SAM) of APTES and PFES, in the hope of creating a p-n junction at the boundary. They found that the intended dopings of the graphene were observed only after the resulting structure was annealed in nitrogen at 200 °C.

A large improvement in the efficiency, to 14.5 %, of the Si-graphene Schottky barrier type of cell described by Miao et al. (2012) has been reported by Shi et al. (2013), who have doped the (CVD grown, transferred) graphene layer with nitric acid instead of a polymer, but find a large improvement when the graphene layer is covered by a ~ 65 nm layer of TiO_2 particles with diameters in the range 3–5 nm. The improvement is nominally from the anti-reflection effect of the TiO_2 layer. The results are summarized in Fig. 3.3. The measured colloid TiO_2 thickness, in the range 50–80 nm, compares well with the optimal antireflection thickness range $d_{TiO2} = \lambda/(4n_{TiO2})$ that is 45–86 nm in the visible range, $\lambda = 400 - 760$ nm, with $n_{TiO2} = 2.2$. (More detail is given in their paper on the anti-reflection estimate.) In their best result the efficiency is reported as 14.5 %, that seems an excellent value. While the main effect of the TiO_2 layer is as an anti-reflection coating, verified by reflection measurements and confirmed by a change in the appearance of the cell, the authors mention that it may have some additional effect in p-doping the graphene.

In their work, Shi et al. (2013) noted that the nitric acid doping could be applied after the TiO_2 film, probably because a network of small cracks appeared in that film upon drying. However, they noted that the benefit of the HNO_3, in their process, decayed with time, although it could be renewed with a second application. This drawback was not noted with the organic doping agent mentioned in the work of Miao et al. (2012).

In summary, the potential for graphene in solar cell applications has been explored only partially. The cells last mentioned, see Fig. 3.3, based on Si single crystals, approach the range of efficiency that is competitive in the market have so-far required the more expensive CVD graphene transferred to the silicon surface. It appears that there are more opportunities to be explored in the area of subsequent annealing and doping direct chemically synthesized graphene flakes deposited to make conductive transparent electrodes. We now turn to other applications of graphene as electrodes, several of which are more advanced than work on solar cells.

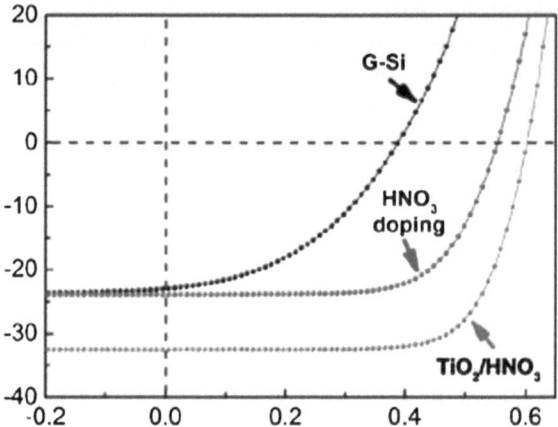

Fig. 3.3 Current density in mA/cm^2 versus voltage in Volts for three types of graphene/Si solar cell, at comparable illumination. The *curve* marked G-Si applies to a Si Schottky barrier cell with an undoped graphene transparent electrode. The *cells* denoted TiO$_2$/HNO$_3$, where the graphene is both doped with nitric acid and covered with the TiO$_2$ anti-reflection layer, are reported as giving 14.5 % conversion efficiency. (Reprinted figure with permission from Shi et al., Fig. 3a. Copyright (2013) by the American Chemical Society)

3.6 Touch Screens and Other Electrodes

Touch screens, transparent conducting electrodes, are used in cell phones, iPads and tablet computers, and store checkouts, among other applications. According to the excellent review of Jo et al. (2012), most of these devices use indium tin oxide (ITO), for which sales are predicted as $3 B in 2014. Graphene is a contender to replace some of the ITO market, because of its superior flexibility and the rising price of ITO, due to the scarcity and high cost of indium metal. The work of Bae et al. (2010) based on CVD deposition of graphene on copper foil, was described in Sect. 2.4. Bae et al. describe steps in producing touch screens for electronic devices using their graphene/PET (188 μm thick polyethylene terephthalate) transparent electrodes. In typical cases, as described in Sect. 2.4, the procedure is iterated to give a stack of four graphene layers on the wide-area PET supporting sheet, and with diagonal dimension as large as 30 in. This stack is a high quality transparent flexible conductor with, at best, ∼90 % optical transmission and low resistance of ∼30 Ω/square. The resulting transparent conductors display extraordinary flexibility, and it was stated that the electrode functions up to 6 % strain.

Supercapacitors (also known as ultracapacitors, or electrochemical capacitors) store electrical charge by forming a double layer of electrolyte ions on high-surface-area conducting materials. The high surface-area provides capacitance values up to 200 F/g, when the material is in an electrolyte, typically in a two-electrode electrolytic cell with a paper membrane separator. The applied voltage is limited to several volts, but supercapacitors can exhibit much higher charge/

discharge rates than batteries, and are thus characterized by high power density. They can be used in applications such as electric cars, where charging can be achieved from the braking process, thereafter to provide extra power for short intervals of time. State-of-the-art supercapacitors are typically based on porous activated carbon (AC), a mass-produced product mentioned in Sect. 2.3.

Supercapacitor devices based on inexpensive chemically exfoliated graphene have been described by Stoller et al. (2008) and by Liu et al. (2010). The latter authors describe high energy density: 85.6 Wh/kg at room temperature and 136 Wh/kg at 80 °C, at a current density of 1 A/g. These high energy densities are comparable to Ni metal hydride batteries, but the charging/discharging times are favorably reduced to seconds or minutes, according to the authors. Zhu et al. describe a commercially scalable process starting from graphite oxide that is exfoliated in a microwave heating step resulting finally in a porous material stated as having a surface area of 3,100 m^2/g. They project a power density of 75 kW/kg for a packaged cell based on their materials.

Hiraoka et al. describe supercapacitor electrodes made from carbon nanotubes. We noted above at the end of Chap. 2 the relatively large production capacity of nanotubes as stated by Segal (2009).

Another potentially large-scale application for graphene is in improving the properties of lithium-ion batteries. In this application graphite and $LiCoO_2$ are typical materials for the anode, where Li ions are reversibly stored in intercalated locations. The charge capacity of the battery is related to the accessible surface area, and the power and charging rates are related to the rates of ion motion, into, and out-of, the storage locations. Relating to graphene, the graphite intercalation compound C_6Li is a benchmark: the charge capacity of this stoichiometry corresponds to 372 mAh/g. Higher performance carbon-based anode materials, with a higher Li-ion accommodation number than in C_6Li, have been studied by Yang et al. (2010), while Yoo et al. (2008) investigated "graphene nano-sheet families" for use in rechargeable lithium ion batteries. The paper of Yoo et al. presents experimental results comparing reversible charge–discharge capacities of anodes: crystalline graphite (measured as 320 mAh/g), GNS (Graphene nanosheet, 6–15 layers thick) (measured as 540 mAh/g), GNS + carbon nanotubes (measured as 730 mAh/g), and GNS + C_{60} (measured as 784 mAh/g). It appears that insertion of nanotubes or C_{60} molecules into the nano-sheet structures opens them to wider interplanar spacing, making more room for the Li ions, whose radius is estimated as 0.06 nm. The authors suggest that the limiting capacity might correspond to the LiC_2 stoichiometry that would evidently correspond to 1,116 mAh/g. The nano-sheet structures are chemically exfoliated, and it is found that enhanced storage capacity correlates with an increased average interlayer spacing, from 0.34 nm, in graphite, to nearly 0.4 nm in the GNS + C_{60} material.

Chemically-derived graphene as a transparent conductor for light-emitting diodes (LED) has been assessed by Wu et al. (2010), in an article containing many useful references. These authors demonstrate organic light-emitting diodes built on chemically-processed graphene that perform comparably to conventional devices using ITO indium tin oxide. Jo et al. (2010) have similarly shown that multi-layer

CVD-grown and transferred graphene serves well as transparent conducting electrodes for GaN light-emitting diodes.

As reviewed by Jo et al. (2012), light-emitting diodes (LED) have large application beyond flashlights and traffic signals, e.g., as backlights for LCD flat-screen television displays. The LCD television market is huge and expected to increase, and LED backlights are expected to be used in 50 % or more of the total number of TV units. The LED backlights conventionally use ITO as the transparent electrode, but ITO is vulnerable, by its high and rising cost, for displacement by a graphene as transparent electrode. In addition, graphene-based LED's are suitable for flexible displays. In LCD displays without backlight, again, the role of the conventional ITO transparent electrode in supplying an electric field across the LCD layer can be supplanted by a graphene electrode.

In a similar fashion, graphene electrodes are applicable in a class of organic field-effect transistor (OFET) devices, see for example Lee et al. (2011). A typical OFET device has a pentacene channel and a gold (Au) gate electrode. The Au electrode can be replaced to advantage with graphene. The OFET devices are inexpensive, and suitable for completely flexible and possibly transparent electronic devices. A good review of this area, with many references, is also supplied by Jo et al. (2012).

In a different class of applications, surface area is needed without electrical continuity, such as possible hydrogen storage by surface adsorption. Monolayer graphene offers the largest surface area per gram of any possible substrate. The potential of few-layer graphene for hydrogen storage has been described recently (Subrahmanyam et al. 2011). These authors note that hydrogenated graphene containing ~ 5 wt % hydrogen is stable and can be stored over long periods. Spectroscopic studies of such samples reveal the presence of sp^3 C–H bonds in the hydrogenated graphenes. The hydrogenated graphenes, suitably, decompose readily on heating to 500 °C, releasing all of the hydrogen, thereby demonstrating the use of few-layer graphene for chemical storage of hydrogen. It was found that the release of hydrogen starts at 200 °C and is complete by 500 °C. The applications of such hydrogen storage are potentially large.

The problem of intermittency in the solar cell production of energy can be alleviated by storage of energy, that might be locally accomplished by using the energy to decompose water, resulting in hydrogen. Hydrogen is itself a valuable commodity, that can be sold on an open market. Its primary commercial uses are in production of ammonia and in the production of gasoline from crude oil. Hydrogen is a portable storage medium for energy, and the chance to store this energy until it is needed is offered by graphene, as explained by Subrahmanyam et al. (2011). The conventional approaches to transportation of hydrogen (as an energy medium) are via pipelines or in cylinders of high pressure gas or of cryogenic liquid. The promising alternative approach to transporting hydrogen is in sorption cells as described.

The authors used a Birch reduction process to add hydrogen to the pristine graphene. The actual process involved lithium in liquid ammonia, and thus is not practical for reversible storage as in the context of an auto. It was not shown how such a process could be adapted to conveniently add hydrogen to graphene, as would be needed in a potential automotive application. Once added, the hydrogen release was shown to be convenient: simply by heating the hydrogenated graphene.

Chapter 4
Graphene Logic Devices and Moore's Law

The primary semiconductor device is the field-effect transistor that has evolved with Moore's Law and is now produced in large scale, reported as 10^{18} per year. These devices have a gate-electrode that draws carriers to a channel connecting the source and drain electrodes, in the ON condition of the device. Graphene has good conduction properties suitable for the channel of such devices, but lacks a high resistance state needed to turn the device OFF. The primary application of FET devices, that has revolutionized information technology through the Moore's Law progression to billion-transistor chips, is as logic and memory devices that function as switches. The physics of the FET switch based on silicon, with gate lengths in the vicinity of 22 nm, no longer allows much reduction in scale. One step has already been taken to reach the present state, a change from naturally-grown SiO_2 as gate insulator to HfO_2 with a larger permittivity, allowing a thick insulator to still attract sufficient charge to the channel for conduction, and avoiding gate leakage by electron tunneling. However, with channel length 22 nm, further reduction is thwarted by "short channel effects" in silicon that appear when the channel depth, into the silicon, of the conducting layer exceeds the channel length or the gate length. To make a metallic conducting contact, as for source and drain, entails an irreducible thickness of the channel that increasingly brings in the undesired "short channel effects". This fundamental difficulty can be overcome with a graphene channel, because the highly conductive monolayer of graphene is only 0.34 nm thick and "short channel effects" are not a problem. In principle, graphene will allow an extension of the transistor packing density beyond that available in silicon. It appears that graphene is unique in this possibility, simply because of its unique position as the thinnest possible excellent electronic conductor. However, there are many difficulties facing this potentially important application of graphene. Because of the potential importance, we explore the various possibilities that might allow a family of graphene-based switching transistors beyond the limit of Moore's Law.

E. L. Wolf, *Applications of Graphene*, SpringerBriefs in Materials,
DOI: 10.1007/978-3-319-03946-6_4, © Edward L. Wolf 2014

4.1 Field Effect Transistor FET Switches

The first difficulty in applying graphene for an FET device comes from the fact that the basic graphene FET device, composed of gapless graphene, essentially cannot be turned OFF, as required for a switch (see the review of Schwierz (2010)). The ratio of On to Off current is desired in the range 10^4–10^7 by logic circuit designers. (One cannot entirely rule out conventional graphene FET devices, since Wang et al. (2008a) have reported large On/Off ratios in sub-10-nm-wide graphene nanoribbon (GNR) field effect transistors. Unfortunately, their method of GNR production is not manufacturable on the mass scale.)

The conventionally configured FET devices described above using a graphene channel suffer from the metallic nature of the graphene, so that a high resistance off-state cannot be achieved. This makes such graphene FET devices unsuitable for logic applications.

A different approach to graphene FETs, with vertical tunnel current flow between parallel graphene layers acting as source and drain, has been recently presented by the Manchester group (Britnell et al. 2012). Several devices, with accurate modeling, are constructed with single crystal layers of graphene and hexagonal boron nitride, with additional devices using MoS_2, instead of BN, as the few-layer mechanically-exfoliated tunnel barrier. The source and drain are separated by this tunnel barrier, the whole sandwich placed horizontally on the oxidized Si wafer that serves as gate electrode. This FET device does not have a conventional "channel", but the forward device current flows vertically by tunneling between the source and drain. The gate voltage modifies the carrier concentrations in the graphene layers, and also, to some degree, modifies the tunneling probability across the single crystal BN or MoS_2 tunnel barrier. It has, remarkably, been possible to create nearly ideal tunnel barriers and tunneling I(V) curves between two graphene monolayers obtained by micromechanical cleavage.

The devices of Britnell et al. sketched in Fig. 4.1, where the graphene layers are horizontal with vertical current flow, can be described as operating by electrostatic doping of the source and drain electrodes.

The diagrams in Fig. 4.1 show the source/drain graphene layers separated by a single-crystal h-BN few-layer tunnel barrier. The isolation or encapsulation of the active graphene layers by h-BN (Dean et al. 2010) largely avoids the spatial electrostatic fluctuations from SiO_2 (that would lead to charge "puddles") and allow nearly ideal tunneling characteristics. This is an unusual accomplishment indeed, but at considerable cost in complexity of manufacture.

Britnell et al. (2012) model the electrostatic doping, using F for electric field, and subscripts b and t to designate bottom and top graphene layers (see Fig. 4.1), with subscript g for gate, and μ as chemical potential (Fermi energy). Britnell et al. give

$$F_b - F_g = n_b e / \kappa \varepsilon_0 \qquad (4.1)$$

Fig. 4.1 Sketches of the Britnell et al. (2012) vertical field-effect tunneling transistor. Layers, from *left* to *right*, are: doped Si, SiO$_2$, BN, Graphene $_{(bottom)}$, h-BN (4 monolayers tunnel barrier), Graphene$_{(top)}$, h-BN encapsulant. (*Top panel*) Device with no bias applied. (*Middle panel*) Positive gate bias to silicon substrate, attracts carriers into both graphene layers (source and drain). (*Lower panel*) Source to drain bias added. The device will operate if graphene$_{(top)}$ is replaced by a metal film. The central insulator, here hexagonal BN with variable number of monolayers, can be replaced with single crystal layers of MoS$_2$ (Reprinted figure with permission from Britnell et al., Fig. 1b, c, d. Copyright (2012) by Science)

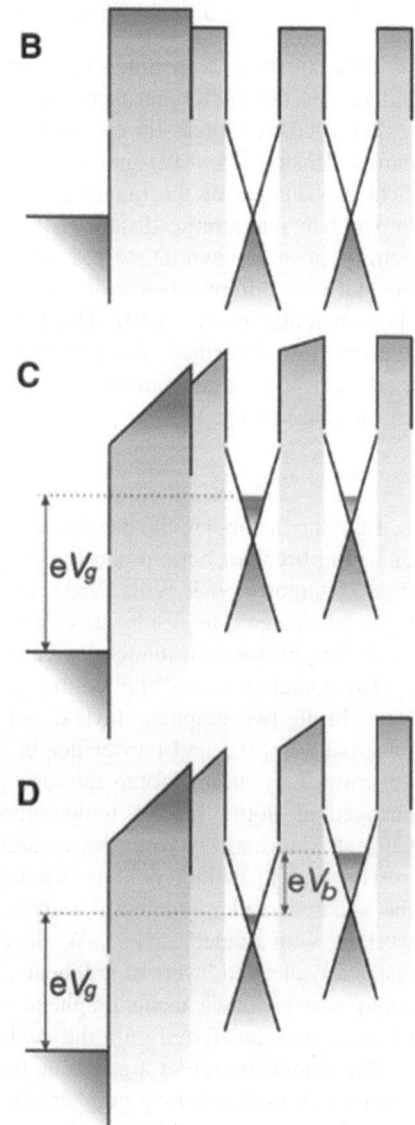

$$- F_b = n_t e / \kappa \varepsilon_0 \qquad (4.2)$$

With voltage V_b between the two graphene layers, one has

$$e V_b = e\, F_b d - \mu(n_T) + \mu(n_B) \qquad (4.3)$$

Combining the relations gives the electrostatic doping condition

$$n_t e^2 d / \kappa \varepsilon_0 + \mu(n_t) + \mu(n_t + \kappa \varepsilon_0 F_g / e) + e V_b = 0. \qquad (4.4)$$

This equation determines the carrier density n_t in the top graphene layer induced by the field-effect from the gate voltage V_G that sets the field F_g.

For a conventional 2D electron gas the Fermi energy is proportional to the carrier density n, and the first term in the previous equation, describing the classical capacitance of the tunnel barrier, is dominant for any realistic spacing d, larger than interatomic distances. In graphene, on the other hand, with its low density of states and Dirac-like spectrum, one finds $\mu(n) \propto \sqrt{n}$, leading to a qualitatively different behavior, that can be described as a quantum capacitance (Ponomarenko et al. 2010). Using the carrier densities as obtained from the analysis just described, the forward (vertical) tunnel current in the device is modeled using the conventional assumption for the electron tunneling conductance σ^T (at gate voltage V_G):

$$\sigma^T \propto g_{bottom}(V_G)\, g_{top}(V_G)\, T(V_G). \qquad (4.5)$$

Here the factors are the densities of states g (see above) and the probability T of tunneling between bottom and top graphene layers at the appropriate bias conditions (Simmons 1963; Wolf 1985). The forward current is obtained by integrating the conductance. The results are close to observation, as shown in inset to Fig. 4.2, an accomplishment in tunnel device construction.

The tunneling theory applied in Fig. 4.2 takes into account the linear density of states in the two graphene layers, and assumes predominant hole tunneling with effective mass 0.5 and barrier height $\Delta = 1.5$ eV. Tunnel barrier thickness d is determined by atomic force microscopy. On–off current ratio, near 50, is little changed at liquid helium temperature, consistent with tunneling mechanism. Current is normalized from device area, in the range 10–100 μm^2.

The work of Britnell et al. is striking in the high quality of the tunnel junctions, the apparent lack of defects, such as pinholes, that are common problems in working with tunnel barriers. While this work was carried out using micromechanically cleaved layers of graphene and h-BN, it seems possible that the devices could also be made using graphene and h-BN grown on Cu, as mentioned in Chap. 2, and transferred onto the device substrate.

The demonstration of a graphene tunneling-based switching device encourages one to seek further among graphene devices for a contender for going beyond the Moore's Law era in CMOS devices. A simpler geometry than that of the Britnell device would be desired to allow large scale production. Generally, tunneling FET devices (TFET) have potential advantages over Silicon FET devices, in lower power consumption.

In order to assess the possibility of graphene tunneling based devices as a replacement for the existing CMOS technology, clearly at the end of its miniaturization range, one needs to understand the characteristics that are required for switching devices in the existing, extremely successful, technology. A sketch of a field effect transistor device is shown in Fig. 4.3, in the N-FET form.

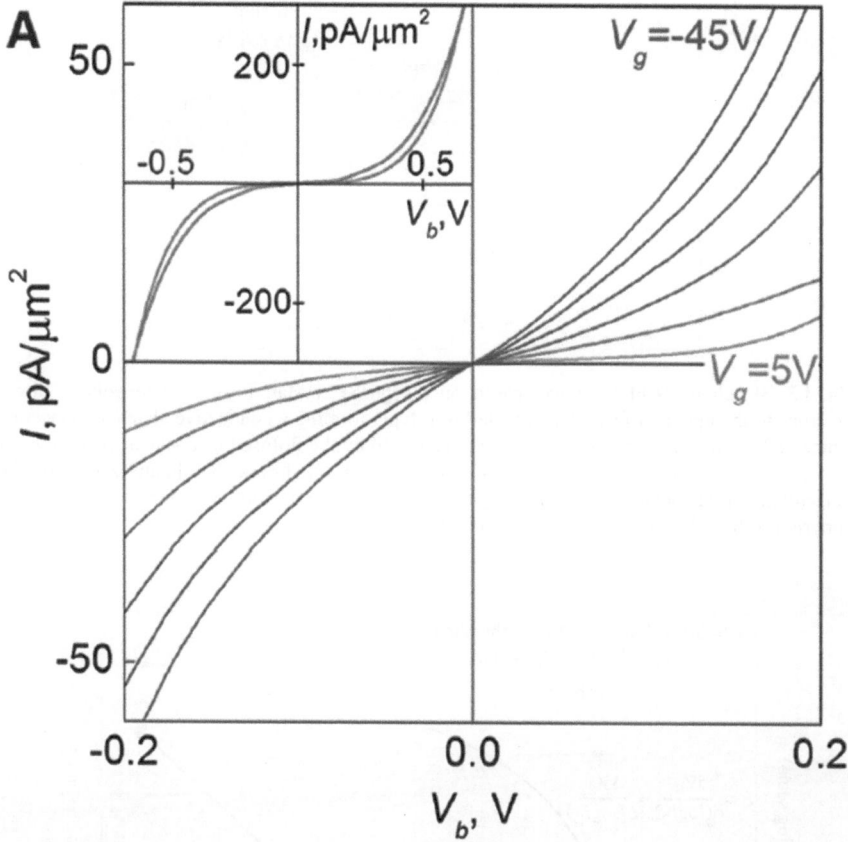

Fig. 4.2 Tunneling characteristics at 240 K for a graphene-(h-BN) transistor device with 6 ± 1 layers of h-BN as the tunnel barrier. I–V curves for different gate voltages V_G in 10-V steps. Because of finite doping, the minimum tunneling conductivity is achieved with $V_G = 3$ V. The *inset* compares the experimental I–V at $V_G = 5$ V (the *lowest curve* in main figure, *upper curve* in *inset*) with theory. (Reprinted figure with permission from Britnell et al., Fig. 3a. Copyright (2012) by Science)

The FET device operation depends on control of the channel conductivity, and thus the drain current, by the gate voltage, applied between gate and source. A positive gate voltage inverts the conductivity type of the silicon in the channel directly under the gate from p-type to n-type. The length of the channel in such silicon devices is now in the range of 20 nm, and exemplary devices have been demonstrated with channel as short as 5 nm. Suitability for logic application centers on how sharply the source-drain current increases as the gate voltage exceeds the threshold value. Aspects of the switching performance of the FET, are shown in Fig. 4.4.

The steepness S of the (log I)/V_g plot is related to the switching (digital logic) capability of the device. In the nomenclature of logic devices, a small value of the "inverse sub-threshold slope S = (d log I_D/dV_{gs})$^{-1}$" (also referred to as SS), is a

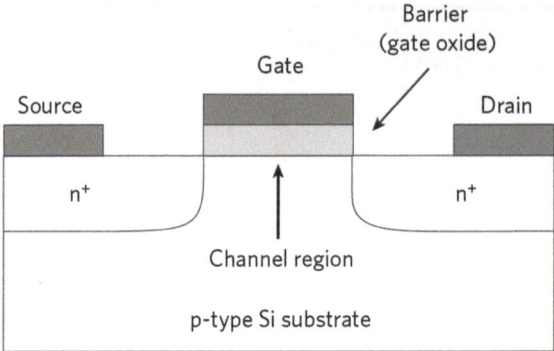

Fig. 4.3 Sketch of N-FET device, based on p-type Si crystal. Positive gate voltage draws electrons to the upper surface, thus inverted to n-type, forming a conductive channel connecting source and drain. The vertical dimension is influenced by the diffused n⁺ contacts, much thicker than graphene, at least 10–15 nm, constituting one of the limits to miniaturization of the conventional FET device. (Reprinted figure with permission from Schwierz et al., Fig. 2a. Copyright (2010) by the Nature Nanotechnology)

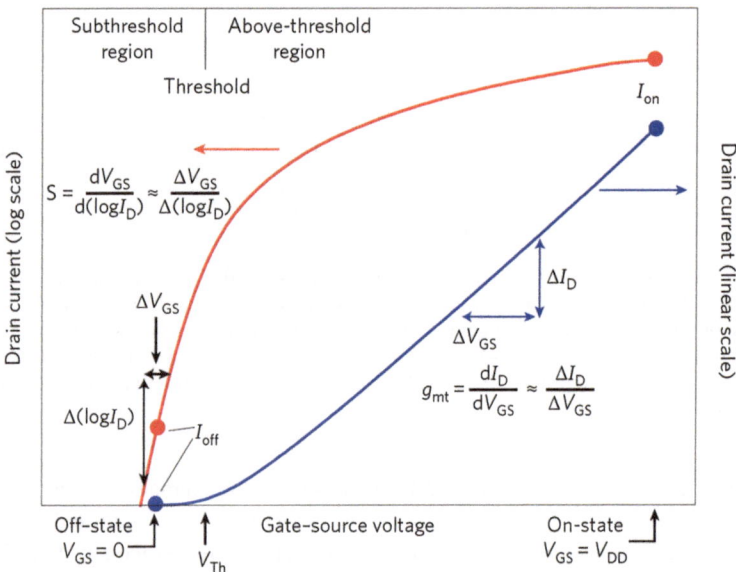

Fig. 4.4 Electrical switching of the field effect transistor FET. Schematic of drain current (log scale on *left*, linear scale on *right*) versus gate voltage. In linear scale, drain current appears at threshold gate voltage, V_{th}, and rises to full On-state value at maximum value of gate voltage, V_{DD}. A figure of merit is the linear slope above threshold, $g_t = dI_d/dV_g$. A second figure of merit, defined on the logarithmic scale, is the inverse sub-threshold slope $S = (d \log I_D/dV_{gs})^{-1}$. A small value of S, ideally 60 mV/decade at room temperature, leading to a steep curve on the logarithmic scale, is considered important. (Reprinted figure with permission from Schwierz et al., Fig. 2b. Copyright (2010) by the Nature Nanotechnology)

figure of merit (see Fig. 4.4). The ideal value for S for a conventional MOSFET is $k_B T \ln 10 = 59.6$ mV/decade at room temperature. As we will see, smaller values can be obtained if the transport mechanism is changed from thermal injection to tunneling. For high speed applications, the device should respond quickly to change in V_{gs}, requiring short gates and fast carriers in the channel. Devices with short channels (the working length will soon be 22 nm) suffer from degraded electro-statics and other problems, collectively referred to as "short channel effects". The scaling theory (Frank et al. 1998; Ferain et al. 2011) predicts that an FET with a thin gate-oxide barrier and a thin gate-controlled region (measured in the vertical direction in Fig. 4.3) will avoid the short channel effects down to very short gate lengths (measured horizontally in Fig. 4.3), perhaps as short as 12 nm. Therefore, according to Schwierz (2010), the possibility of having a channel just one atom thick is one of the most attractive features of graphene, in possible application to FET switching transistors. If channels are made thin in the silicon technology, surface roughness is usually introduced, that reduces mobility or even makes an open circuit, but such an effect is absent using graphene, that is both thin and smooth. In CMOS logic, a large number of devices are always in the "off" state, yet drawing some power. Low current in the "off" state, corresponding to a large on/off current ratio, is desired, and the expectation is that a successful device family will have a ratio at least 10^4. This leads to the estimate, for a conventional semicon-ductor FET, that the band gap should be at least 0.4 eV. Further, since the devices are used in complementary P-FET/N-FET pairs, it is desired that the P- and N-devices shall have symmetrical properties. This condition is well met by graphene, whose energy bands are symmetrical about the neutral point, while the substantial difference in electron- and hole-effective masses in conventional semiconductors has required adjustments in design to restore operational symmetry.

Silicon research FET devices have been demonstrated at gate lengths as small as 5 nm (Wakabayashi et al., 2006). But clearly there is a limit, and soon the scaling progression, at least in Si logic devices, will stop. A set of final steps in the Moore's Law scaling, based on multi-gated Si-on-insulator (SOI) devices, is given by Balestra (2010), in an article entitled "Silicon-Based Devices and Materials for Nanoscale CMOS and Beyond-CMOS". A similar position is arrived at by Ferain et al. (2011), who predict that multi-gated silicon devices will reach the 3 nm node (the basic linewidth to be 3 nm), and that this will take about 20 years. (These authors are certain that devices in 20 years time will be silicon multi-gated FETs.)

A second aspect, however, is the power dissipation, that continues to rise in the projected remaining steps of the Moore's Law progression. It has been established (Wang et al. 2004; Ionescu and Riel 2011, and work cited therein) that *tunneling* FET devices (T-FET) can lower the power dissipation by a factor perhaps 100 from the Si CMOS family of devices. For suitably small off-state current I_{off}, the energy usage is closely related to the energy per switching-operation. Switching-energy scales with the supply voltage, V_{dd}, as

$$P \propto I_{off} V_{dd}^3 \qquad (4.6)$$

According to Ionescu and Riel (2011), the tunnel FET could allow the supply voltage V_{dd} to reduce from 1 to 0.2 V, giving a reduction in power of 125. It appears that T-FET devices can be made in several different materials, including Si, III-V compounds like GaAs, and carbon (nanotube or graphene), and it may well be that the conventional materials would be easier to incorporate into the silicon technology. Before considering possible graphene tunneling logic, we give an indication of where the Si CMOS devices stand, as suggested by the work of Wakabayashi et al. (2006).

Using several fabrication innovations within the silicon technology, these authors produced a family of MOSFET research devices with channel lengths no more than 5 nm, and suggested that they operate in a useful fashion. The drive-currents reached are on the order of 0.1 mA/μm. On the other hand, the drain currents do not saturate at high drain voltage, but are linear to 0.4 V. These are non-ideal effects but may not remove the utility of the device.

Wakabayashi et al. (2006) show (their Fig. 2) a Transmission Electron Microscope (TEM) image of gate electrodes with 5 nm gate length on the Silicon substrate of their devices. The striking aspect is that the vertical extent of the gate structure is at least ten times larger than the length of the channel. This strongly violates the condition of strong electrostatic control of the channel, to avoid "short channel effects". The gate electrode is polysilicon, reduced in width at its lower end, and enclosed by a Sidewall layer of Silicon Nitride and by a Silicon dioxide liner. The device also involves $CoSi_2$ silicide source and drain contacts, augmented with "shallow source and drain extensions" to contact the extremely short channel. The vertical extent of these structures exceed the channel length.

The scaling rules of Frank et al. (1998) and Ferain et al. (2011) favor thin structures of minimum vertical extent (perpendicular to the channel). The structures of Wakabayashi et al. (2006) violate this criterion, and further miniaturization of their design will make the violation larger.

To return to graphene FET devices for switching applications, the conventional route to avoid the metallic nature of graphene, that prohibits a conventional FET that will turn off, is to use a graphene nanoribbon (GNR) in which a bandgap does appear, from size quantization. The difficulty is that to achieve a suitably large bandgap, at least 0.4 eV, the ribbon has to be extremely narrow, less than 10 nm wide. Such ribbons have been selected from suspensions resulting from sonication of graphite flakes, in the work of Hongjie Dai and collaborators (Li et al. 2008) yielding FET's of with On/Off ratios on the order of 10^6. From the point of view of logic FET devices, these ribbons are better than carbon nanotubes, in that they do reliably have energy gaps, but they suffer from the same problem as do carbon nanotubes. Namely, there is no production method of getting the right size GNR or nanotube to the right location on the chip, that in current practice may contain a billion FET transistors in a centimeter square.

This forces a conclusion that a realistic device will have to be deposited and patterned on the planar chip surface, there being no prospect of importing, and accurately locating, vast numbers of selected nanoribbons. On the other hand, present and future lithography does not permit patterning 10 nm wide nanorib-bons, that, further, would need atomically accurate edges. A possible solution to the problem of rough edges on nanoribbons may be offered by the chemical route to nanoribbons described in the work of Cai et al. (2010) and Koch et al. (2012). These methods do clearly offer atomically precise edges for narrow nanoribbons, but leave open the difficulty of importing and locating the graphene elements.

These points, with a second possible exception that we will shortly mention, *exclude* GNR devices, within a logical point of view based on mass manufactu-rability needed in the context of extending Moore's Law.

The second possible exception, to the non-manufacturable assessment of graphene nanoribbon devices, is recently suggested in the works of Sprinkle et al. (2010) and Hicks et al. (2012). Their innovation is to specify, on an atomic size scale, the widths of nanoribbons (and also to specify the edge configuration as armchair or zigzag) by growing the graphene on etched groove geometries pre-patterned on SiC. The latter paper is entitled "A wide-bandgap metal–semicon-ductor-metal nanostructure made entirely from graphene". In their nanostructure arrays of 12- or 18 nm-deep trenches are patterned into SiC, onto which about 400 parallel monolayer graphene ribbons of length 50 μm, and widths 15 and 36 nm, are then grown. The ribbons are centered on the trench sidewalls, using methods described in the cited works of Sprinkle and Hicks. The sidewalls of the Hicks et al. (2012) trenches are (2207) and (1103) facets of SiC, with angles from the vertical (normal to the (0001) surface) about 30°. As shown by Sprinkle et al. (2010), an initially vertical sidewall will reconstruct into a sloping (110n) facet if the SiC is heated to 1,250 °C, and, further, a monolayer of graphene will grow on that sidewall if the crystal is further briefly heated to 1,450 °C. Sprinkle et al. (2010) made an array of 40,000 transistor devices per cm^2 on SiC (0001) using this process. Hicks et al. give evidence that the sharp curvature at the join of the top (0001) plane and the sloping trench wall makes a 1.4 nm-wide semiconducting graphene nanoribbon. This ribbon is epitaxially connected to an n-type metallic nanoribbon on the (0001) top surface and a p-type conducting nanoribbon on the sloping facet wall. The result is a narrow nanoribbon whose armchair edges are aligned along the corner between the (0001) top and sloping (110n) sidewall surfaces. Strong experimental evidence of the bandgap, restricted to the 1.4 nm-wide strip at the sidewall intersection, was provided by angle-resolved photo-emission spectroscopy (ARPES) measurements. These epitaxial structures are of atomically controlled dimensions and atomically specified edge geometry, and are single-crystalline (via epitaxy to the SiC) over lengths of at least 50 μm. This class of nanoribbons, extending to nanostructures with semiconductor–metal junctions, seems a candidate for a dense FET device technology based on graphene nano-ribbons (Berger et al. 2004).

Beyond the SiC option just mentioned, the further remaining options for graphene dense logic devices therefore lie with wide area graphene devices, that do not require a narrow width to obtain a gap. Three different types of such wide area, manufacturable, graphene devices have been proposed. These are (1) based on an electrically induced gap in bilayer graphene, in a traditional FET

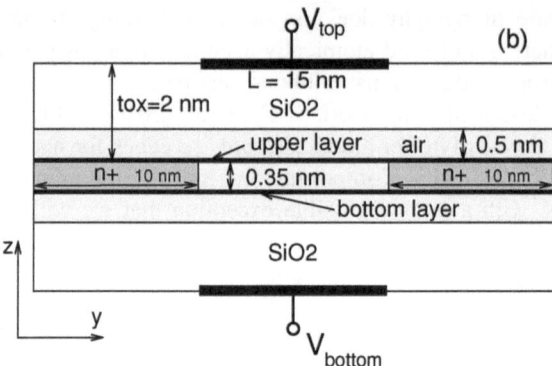

Fig. 4.5 Tunable-gap bilayer Graphene FET as modeled by Fiori and Iannaccone (2009a). The vertical height of the device as shown is 4.35 nm, excluding the width of *top* and *bottom* metal (or graphene) gate electrodes. The overall height is small compared to the channel length, shown as 15 nm. In this figure, bilayer graphene is represented as two single layers of graphene separated by 0.35 nm. It is assumed that source and drain are bilayer graphene, heavily doped to achieve strong n-type conductivity. (Reprinted figure with permission from Fiori and Iannaccone et al., Fig. 1b. Copyright (2009a) by the IEEE)

configuration; (2) a bilayer graphene tunnel FET device, and (3) a purely two-dimensional graphene-based tunnel transistor. The second two devices are based on interband tunneling. We provide some review of that topic after mentioning the first graphene device, of the conventional FET form.

The first graphene FET design that we describe is that modeled by Fiori and Iannaccone (2009a). This device uses bilayer graphene, with a vertical electric field, applied by top and bottom gate electrodes, to induce a bandgap (McCann 2006). The device, called the Tunable-gap Bilayer Graphene FET, is modeled numerically. As can be seen in Fig. 4.5, this proposed device is thin, about 5 nm in vertical extent.

The overall height of the proposed structure in Fig. 4.5 is much less than the channel length, quite unlike the silicon devices of Wakabayashi et al. (2006), described above. This is an advantage, in general terms, to avoid undesirable short-channel effects. The assumption is that such a structure could be achieved by transferring bilayer graphene, grown on copper foil, for example, onto an insulating substrate, followed by lithographic patterning. The width of the device (out of the page) is not a crucial parameter in this and following designs, and should not conflict with present capabilities in lithographic patterning. Careful modeling by the authors predicted that this device operates, but, with the bandgap available by electric field in bilayer graphene, unfortunately, does not have a large enough On/Off ratio to be suitable for logic applications.

4.2 Tunneling FET Devices

We now turn to tunneling field effect transistor (TFET) devices that operate by interband tunneling. For background on interband tunneling, the Zener diode is a commercial silicon device, a pn junction, that, when put into strong reverse bias (increasing the built-in shift of the bands on the opposite sides of the junction), abruptly becomes highly conductive, as carriers tunnel from the valence band on one side, across the depletion region, to empty states in the conduction band on the opposite side. The physics of this working device was initially explained by Zener (1934). A more recent analysis of the interband Zener tunneling current is given by Kane (1959). This current is independent of temperature, and given by Kane's formula for n, the rate of electron transfer at electric field E, with $F = eE$, for a semiconductor of bandgap E_G, as

$$n = [F^2 m_r^{1/2}/(18\pi\hbar^2 E_G^{1/2})] \exp(-\pi m_r E_G^{3/2}/2\hbar F). \tag{4.7}$$

Here the reduced mass m_r is derived from the electron- and hole-mass values, $m_r = m_e m_h/(m_e + m_h)$. The carrier mass changes in character from electron to hole as it crosses the depletion region. This formula (4.7) strongly favors semi-conductors of small bandgap, and requires a high electric field E.

A related device is the Esaki (1958) diode, where highly degenerate n- and p-layers, with consequent small depletion width, are needed to make the tunneling probability usefully large at zero bias. This device was part of the Nobel Prize in Physics in (1973).

The initial suggestion, and analysis of TFET devices based on interband tunneling, was early given by Quinn et al. (1978), who considered a p-i-n structure. In the Quinn structure, p and n are heavily doped regions on a weakly p-type silicon surface, with a field electrode shifting the bands in the connecting i region (weakly p-type Si) sufficient to allow Zener tunneling into the empty conduction band. The general shape of the energy bands in such a device is shown in Fig. 4.6.

Early reports of operational tunneling TFET devices on silicon were given by Reddick and Amaratunga in (1995) and by Koga and Toriumi (1997).

In an important paper, Wang et al. (2004a) discussed the low power dissipation possibility of the T-FET, and also the faster switching, with a smaller S factor, not limited by $k_B T/e$. Their p-i-n devices are built on a Si surface by conventional diffusion of dopants, and a gate covers the intrinsic i region. The devices are operated in reverse bias of p to n and are described as a "MOS-gated reverse biased p-i-n diode. The band-to-band tunneling can be controlled by the gate voltage and the leakage current is minimized due to the reverse biased p-i-n structure". The authors find that complementary logic is possible using these devices: "complementary NTFET and PTFET similar to NMOS and PMOS are fabricated on the same silicon substrate". These aspects were further discussed, and demonstrated in carbon nanotube T-FET devices, by Appenzeller et al. (2004).

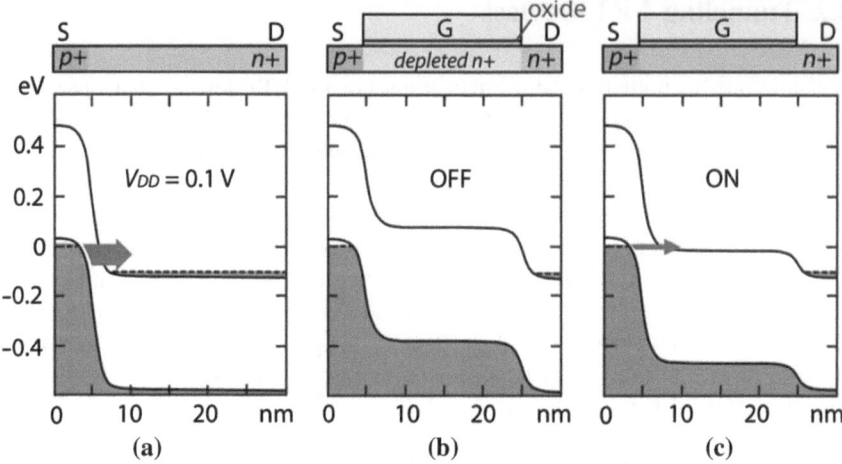

Fig. 4.6 Energy band diagram and layer structure for a tunneling field effect transistor T-FET consisting of a p$^+$ source, an n$^+$ drain and a gate (G). In (**a**) the channel is shown without a gate, and with sufficient source-drain bias $V_{DD} = 0.1$ V to drive Zener tunneling into the conduction band of the drain from the valence band of the source. In (**b**) the gate fully depletes the channel creating the OFF condition of the gated T-FET. In (**c**) a positive gate potential turns the channel ON, with current set by overlap of valence band electrons with unfilled conduction band states. (Reprinted figure with permission from Seabaugh and Zhang et al., Fig. 1. Copyright (2010) by the IEEE)

Knoch et al. in 2007 provided more analysis toward practical T-FET devices, and experimentally found values of the parameter S smaller than 60 mV/decade, with a tradeoff between small S and high ON-current for the device. An extensive review of the subject was given by Seabaugh and Zhang (2010).

4.3 Graphene Tunneling FET Devices

A tunneling field effect transistor T-FET in bilayer graphene is described by Fiori and Iannaccone (2009b). This device (Fig. 4.7) is geometrically similar to the graphene device in Fig. 4.4, but the mode of operation is changed. Now, forward current is by interband tunneling, between n$^+$ source and p$^+$ drain in bilayer graphene. Forward current is enabled at a gate voltage that lines up the valence band and conduction band energies in source and drain electrodes.

The principle illustrated in Figs. 4.6 and 4.7 was demonstrated by Appenzeller et al. (2004), in a device based on a carbon nanotube, with an Aluminum top gate electrode. The conditions in the carbon nanotube are similar to those in the graphene device described by Fiori and Iannaccone (2009b). These authors find, that, with a drain to source voltage of 0.1 V, an On/Off ratio larger than 1,000 can be reached, even within the limitations on achievable bandgap in the bilayer

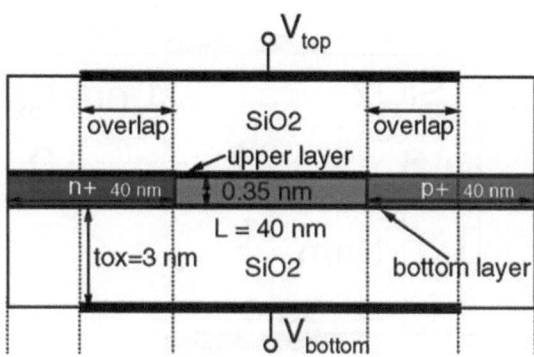

Fig. 4.7 Bilayer graphene tunnel T-FET device. Bilayer graphene is shown as two graphene layers separated by 0.35 nm. In the central region, of width L = 40 nm, the bilayer is undoped and its bandgap is enhanced by electric field, established by the two gates. The main contribution to the shift of band energies between p- and n-electrodes results from the large fractional chemical dopings, on the order of 2.5×10^{-3} that are assumed in the bilayer graphene. "Overlap" in the figure indicates regions where gate extends out over the heavily doped source and drain. (Reprinted figure with permission from Fiori and Iannaccone et al., Fig. 1. Copyright (2009b) by the IEEE)

graphene (in the range of a few hundred millielectron Volts). The switching in this tunneling device is superior, with an S factor (see Fig. 4.4) reported as 20 versus 60 meV/decade for the best thermionic device. In Fig. 4.7, the large vertical field, established by the top and bottom gates, maintains the bilayer band gap in the length between those gates, but the bandgap of the bilayer graphene goes to zero in heavily doped regions (beyond the extent of the gates). The use of SiO_2 dielectric, as shown in total height of 6 nm, permits an electric field up to 14 MV/cm before breakdown. Manufacturing such a device could be done, in principle, by transferring bilayer graphene, grown on Cu foil, onto the device substrate, and then using masking to deposit n- and p-type dopants specifically onto the source and drain sections of the device, under typical CVD conditions. The width of the device (out of the page) is not a crucial parameter, so the suggested device is compatible with present silicon technology. The authors assess this device as promising for fabrication and circuit integration, based on their simulational results.

Finally, a Monolayer graphene double-gated tunneling transistor is described. This second conceptual approach (modeled, but not experimentally demonstrated) to graphene tunneling T-FET transistors with large On/Off ratio (Fig. 4.8), is a single graphene sheet (wide ribbon, including source and drain) with an inserted barrier (Fiori et al. 2011, Fiori et al. 2012). The ends of the wide ribbon are heavily doped P-type, and the narrow barrier is hexagonal BCN. This barrier is envisioned as an inserted BCN film (epitaxed to the graphene channel) with the same hexagonal structure and lattice constant as graphene. (The modeled device is PFET, but there is no basic problem to make the complementary NFET in this form.) Methods that may be adaptable to inserting the tunnel barrier are given by Levendorf et al. (2012). It is pointed out that insertion of the BCN barrier adds flexibility to design of graphene transistors. The BCN barrier region can be

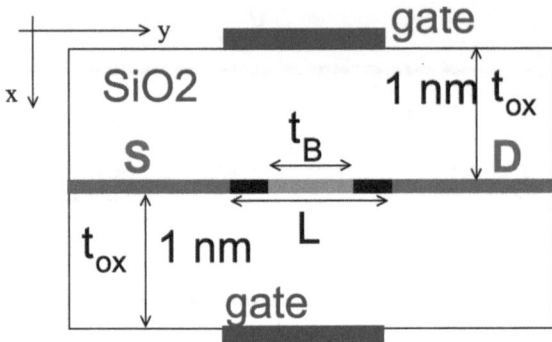

Fig. 4.8 Monolayer-graphene double-gated tunneling transistor with high I_{ON}/I_{OFF} ratio. This proposed device (here PNP, or PFET), is monolithic single layer graphene, modified in the center by chemical change from pure C to BCN (in the same hexagonal lattice, and containing a bandgap), introducing a conventional tunnel barrier. (Reprinted figure with permission from Fiori et al., Fig. 1. Copyright (2012) by the American Chemical Society)

tailored, by choice of composition, to have a barrier ranging from 1 to 5 eV (with some caveat to control large domain formation, vs. homogeneous films), and, further, can be designed to pass one kind of carrier while blocking the other. The structure suggested in Fig. 4.8 is a hybrid h-BCN-graphene device, with a chemically-formed, gate-controlled barrier inserted between Source and Drain. The barrier comprises an epitaxial length t_B of h-BCN. Hexagonal BCN potentially can have variable fractional C content from 0.1 approaching to 1.0, with energy gap adjustable in the range 1–5 eV. The h-BCN can be grown in 2–3 atomic layer films on copper foil by CVD, using methane and ammonia-borane (NH_3–BH_3) as source gases. It may be possible to introduce the h-BCN barrier region starting with pure graphene and masking, so that the desired barrier region is exposed briefly to NH_3–BH_3 under typical CVD conditions. In any case, the graphene in this device is of large area, and fabrication, in principle, does not suffer from difficulty in reducing its width, that would require (in a GNR device) atomically defined edges, unattainable in present or future lithography. The modeled device was of the PFET form, with source and drain both of heavily doped p-type graphene. The simulated performance of the device is shown in Fig. 4.9, with actual modeled barrier composition BC_2N.

The performance of the simulated device is acceptable from the point of view of CMOS logic. The dashed line marked "60 mV/dec" in Fig. 4.8, is a measure of the steepness of the (log I)/V_g plot, related to the switching (digital logic) capability of the device (see Fig. 4.4). The S value, from the points in Fig. 4.9, is about 80 mV/decade, acceptable, and clearly superior to that of the simple graphene FET (upper points in Fig. 4.9).

The design of the TFET graphene logic device in Fig. 4.8 is superior to any possible Silicon device, from the point of view of having a conducting channel of thickness much smaller than its length, because the depletion depth in any silicon

Fig. 4.9 Simulated drain current I_D versus gate voltage, for graphene-h-BC_2N tunnel FET device (lower data points), compared with simple graphene FET transistor (upper points). The tunnel device exhibits I_{ON}/I_{OFF} ratio exceeding 10^4, with source-drain voltage 0.6 V at room temperature. The different symbols in the *lower curve* represent slightly different assumptions on the exact location of the middle of the barrier. The length of the barrier t_B (See Fig. 4.7) is 5 nm, and the doping fraction in the p + leads is 10^{-2}. (Reprinted figure with permission from Fiori et al., Fig. 4. Copyright (2012) by the American Chemical Society)

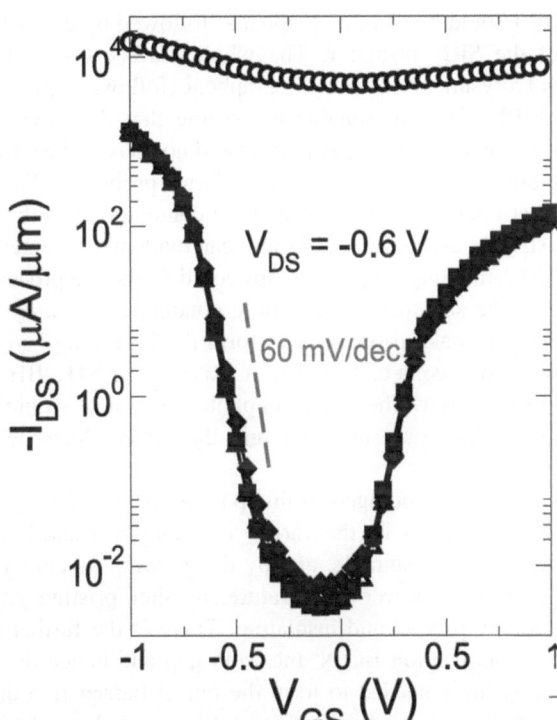

device is large on the scale of the graphene thickness, 0.34 nm. The modeled device in Fig. 4.8 is about 2.5 nm in height, excluding the top and bottom gates. If the top and bottom gates are graphene (graphene as interconnects are discussed below) the total height of the device would be less than 4 nm. The off-current in a tunneling device can be exponentially small, but the forward current may be limited by the conductance quantum e^2/h (≈ 25.813 k$\Omega)^{-1}$. (At $V_{dd} = 0.5$ V, the forward current in that case would be ~ 20 μA.) Polycrystalline silicon (Polysilicon), employed to make gate electrodes in present device technology, is conveniently provided by chemical vapor deposition (CVD). The same technology is adaptable, with limitations described in Chap. 2, to deposit graphene, or BCN or insulating BN, the materials that might replace Silicon. It seems clear that such graphene tunneling devices are capable in principle of extending the scaling past the limit of silicon devices, but the practicality remains to be seen.

4.4 Manufacturability of Graphene Devices

To speculate about manufacturability of devices such as that of Fig. 4.7, using photolithography, their double-gated nature would mean that a large substrate, such as a silicon wafer or glass sheet would be prepared with patterned electrodes (that could be n^+ or p^+ graphene) followed by deposit (or growth from polysilicon) of the SiO_2 insulator. The whole surface could then be covered with a large polycrystal of monolayer graphene following methods like those of Bae et al. (2010). (It is reasonable to assume that the orientation of the graphene lattice relative to the wiring pattern would not need to be specified, and that the occasional grain boundary will not be a large problem.) The patterning of the graphene electrodes would then be done in contemporary lithography, where the resolution, using the ArF laser at 193 nm, can evidently be extended to the vicinity of 20 nm. The remaining graphene strips could further be processed, after patterned masking, to make selected regions more conductive (e.g., by depositing K or Ca, or, e.g., SbF_6 to make p-type), or to form the insulating barrier regions by brief exposure (e.g., to plasma of B and N or exposure to NH_3–BH_3), to form the BCN insulator tunnel barrier. These steps in photolithography would be followed by deposit of the upper SiO_2 gate insulator and the upper electrode structure, with reference to Fig. 4.7.

One is encouraged in this possibility of a family of graphene TFET manufacturable devices by the variety of monolayer and bilayer graphene structures that appear to the suitable, and by the growing inventory of methods to deposit CVD graphene at lower temperature, to alter pristine graphene into strongly n-type, strongly p-type, and insulating. There is the further possibility of controlling, by the composition BC_xN, the band gap and hence the width t_b of a modified insulating layer needed to form the tunnel barrier in a device like that of Fig. 4.8.

As further evidence of flexibility in graphene device design, Fiori et al. (2011) investigated resonant tunneling devices, capable of negative resistance behavior. The simulated resonant tunneling devices were fully planar, similar to the device shown in Fig. 4.8, except that the barrier region is now split into two BN barriers (width t_B) with an intervening narrow graphene region (width w). Localized states are now possible on the interior graphene region of width w: when such states align in energy with the Fermi level of the source, a peak appears in the I_D-V_{GS} characteristic. These authors argue that such a proposed family of resonant tunneling transistors achievable in graphene may be superior to those obtained in III-V materials, because the graphene device offers better electrostatic control of operation.

From the dimensions of the devices in Figs. 4.7 and 4.8, it is not clear that a higher planar packing density can be achieved than is being achieved in silicon. The value of the TFET devices that likely can realized in graphene is the lower power density. Power density is one of the main drawbacks to silicon technology at the end of Moore's Law. Cloud computing, in very large installations, may still change to superconducting logic to save energy (Bunyk et al. 2001). But the high power consumption in tablet and laptop computers and in cell phones, where related short running times between battery charges, could be reduced by adoption of a TFET family of devices as we have discussed, with inherently lower power dissipation levels, that do not need the refrigeration that is practical for a large installation.

Finally, we briefly compare that possibility of graphene FET switches, that we have presented above as potentially manufacturable (but at present scarcely demonstrated), with carbon nanotube FET switches, CNT. The carbon nanotube is a rolled cylinder of graphene, and the problem of irreproducible boundary conditions for a graphene nanoribbon is absent in the formed nanotube. The diameter of the nanotube is $d = a(n^2 + m^2 + mn)^{1/2}$, where $a = 0.249$ nm, and where n, m are integers that describe the angle between the hexagons and the axis of the nanotube. A subset of nanotubes, depending upon n-m, are semiconducting, with energy gap inversely proportional to d. FET transistor devices of high quality have been fabricated and measured by Franklin et al. (2012) on single nanotubes of diameter about 1.3 nm with reported bandgap about 0.62 eV. The high bandgap allows a conventional FET with ON/OFF current ratio exceeding 10^4, even with a channel length as small as 9 nm. The problem of manufacturability now appears in the need to select only those nanotubes that exhibit an energy gap, from a CVD grown array containing many metallic tubes. This manufacturability question has been addressed very recently by Shulaker et al. (2013), who make a parallel array of nanotubes with common source, drain and gate electrodes, and destroy the metallic subset by applying a voltage pulse between source and drain. In an aggressive fashion these workers have demonstrated a working computer made only of carbon nanotubes.

Chapter 5
Niche Applications of Graphene Within Silicon Technology

In Chaps. 3, 4 we have outlined possible applications of graphene in two leading potential areas, solar cells and switching logic devices. These are arguably the largest applications that graphene might find, but are both only moderately likely to occur with large impact. More certain are niche applications within the existing silicon technology, as well as earlier mentioned applications as additives, and as transparent electrodes in a variety of areas including touch screens as in cell phones and consumer check-out stations. The niche applications in semiconductor technology include radio frequency transistors, chip interconnects, and flash memory.

5.1 High Frequency FET Graphene Transistors

We first focus on graphene FET high-frequency devices, that definitely are contenders in a post-silicon era for rf amplifiers. Such devices do not need to have a fully off state, so the essentially semi-metallic nature of graphene is not prohibitive. At the time of the Schwierz (2010) review the fastest graphene FET had a cutoff frequency of 100 GHz with a gate length 240 nm (Lin et al. 2010). A sketch of the record-breaking device is shown in Fig. 5.1. To reach the Dirac neutral point of the film (grown on SiC) always required a highly negative gate voltage $V_G < -$ 3.5 V, indicating a highly conductive n-type metal in the channel at zero bias voltage, $n > 4 \times 10^{12}$ cm^{-2}. The authors point out that this is advantageous to obtain a low series resistance in the device. The cutoff frequency is defined as the highest frequency where current gain is provided by the device, as shown in Fig. 5.2. The graph shows that 100 GHz can be reached with a gate length of 240 nm.

Referring to Fig. 5.1, the graphene was grown on the Si face of semi-insulating high purity SiC by thermal annealing at 1,450 °C, and exhibited an electron carrier density $\sim 3 \times 10^{12}$ cm^{-2} and a Hall-effect mobility in the range 0.1–0.15 m^2/Vs. The source and drain are long, closely-spaced parallel electrodes formed by sequential deposits of Ti (1 nm) and Pd (20 nm) followed by Au (40 nm), using

E. L. Wolf, *Applications of Graphene*, SpringerBriefs in Materials,
DOI: 10.1007/978-3-319-03946-6_5, © Edward L. Wolf 2014

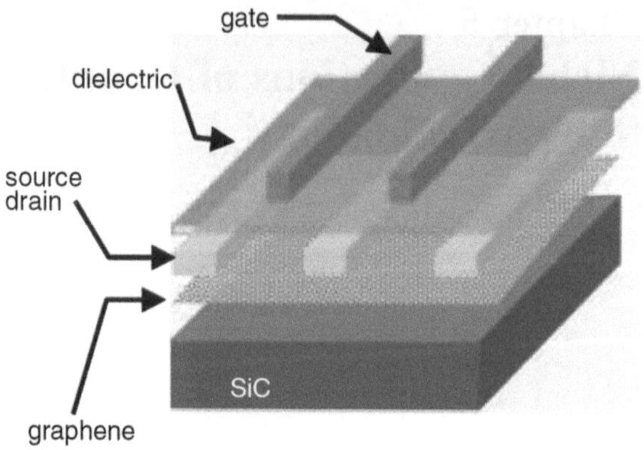

Fig. 5.1 Sketch of graphene field effect transistor FET fabricated on a 2-in. SiC wafer. Graphene grown on SiC tends to be inherently conductive by electrons. The source and drain are shown, and the *top* gate electrode. (Reprinted figure with permission from Lin et al., Fig. 1a. Copyright (2010) by the Science)

Fig. 5.2 Current gain for graphene FET as sketched in Fig. 5.1, as function of frequency. (Reprinted figure with permission from Lin et al., Fig. 1d. Copyright (2010) by the Science)

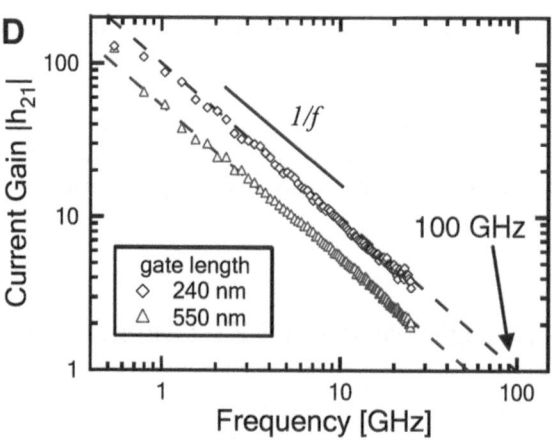

e-beam lithography. To define the graphene channel, and to isolate individual devices, regions of graphene were etched in an oxygen plasma using PMMA as etch mask. To form the gate electrode, an interfacial polymer layer, composed of a derivative of poly-hydroxysilane, was spin-coated onto the graphene, before atomic layer deposition (ALD) of 10 nm thick HfO_2. These steps were designed to least degrade the electron mobility in the graphene, that afterwards was in the range 0.09–0.1520 m^2/Vs for devices across the 2-in. SiC wafer. The measured cutoff frequency exceeds those of previously reported (to 2010) graphene FETs, as well as of Si metal–oxide–semiconductor FETs of the same gate length. A summary plot (shown in Fig. 5.8 of Schwierz (2010)) suggests that 1 THz may be available at gate length 20 nm. In his review, Schwierz (2010) suggests that graphene FET devices may achieve superior performance by allowing a shorter channel.

Drawbacks to the design of Lin et al. (2010) include the high temperature needed (1,450 °C), the mobility degradation due to the graphene growth on SiC

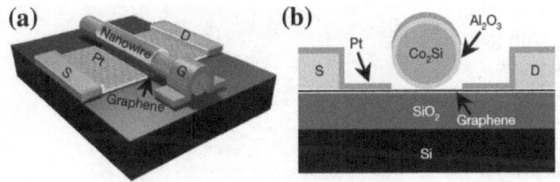

Fig. 5.3 Self-aligned nanowire gate design for high-speed graphene FET. Micro-mechanically cleaved graphene transferred to oxidized Si wafer, evaporated metal source and drain, and Co_2Si-Al_2O_3 core–shell nanowire *top*-gate, covered with self-aligning Pt gate electrode. Diameter of nanowire sets channel length, with 5 nm Al_2O_3 shell functioning as gate dielectric. The self-aligned Pt thin-film pads extend the source and drain electrodes optimally to the location of the gate. (Reprinted figure with permission from Liao et al., Fig. 1. Copyright (2010) by the Nature)

and due to the processing needed for the gate deposition, and the difficulty in precisely aligning the top gate to match the channel. The latter issue is addressed in a device with a self-aligned gate, in the work of Liao et al. (2010).

A radio frequency (RF) FET with narrow channel width, achieved by a self-aligned nanowire gate, has been demonstrated by Liao et al. (2010). This device is not a candidate for production, however, because it is based on a flake of micro-mechanically cleaved graphene transferred to an oxidized silicon wafer, as shown in Fig. 5.3. This preserves the high mobility of the graphene, that may eventually be also achieved with CVD based production scale processes. The present devices achieve a higher cutoff frequency, about 300 GHz, and higher current density, 3.32 mA/μm, compared to about 0.8 mA/μm for the devices of Lin et al. (2010).

Extremely high current densities, on the order of several mA/μm, are reported in these measurements. Assuming the thickness of the graphene is 0.34 nm, the current density is approximately 3×10^8 A/cm^2. A discussion of current density in graphene is given by Efetov and Kim (2010).

DaSilva et al. (2010) have addressed questions of carrier velocity and its saturation in these devices. They find that the primary scattering mechanism at high current density is emission of surface optical phonons of the substrate. It appears that one of the advantages of the diamond-like-carbon substrate, DLC, is that its phonons are at a higher energy, 165 meV, less likely to be excited. High electric field effects on carriers in graphene are also discussed by Tani et al. (2012).

Graphene transistors using ferro-electric gates have been analyzed by Zheng et al. (2010), including a review of different possible dielectrics.

Bilayer graphene FET devices have been fabricated and modeled by Szafranek et al. (2012). In comparison with similar devices with monolayer graphene, they find a voltage gain of 35, a factor six higher than they can attain in monolayer graphene FET devices.

The radio frequency transistors above are based often on SiC-grown graaphene. Growth of graphene on SiC is appealing because wafers of SiC are available and large arrays of graphene devices can be fabricated. The drawbacks are the high temperature needed and the low mobility of graphene grown on SiC. Liao et al.

Fig. 5.4 TEM cross-section of graphene FET, with gate width 40 nm. The oxide thickness is about 20 nm. The graphene layer is invisible in the TEM image, and the diamond-like carbon layer (the base in the schematic diagram, appears to be about 50 nm thick. The TEM image suggests that this design cannot be made much smaller. (Reprinted figure with permission from Wu et al., Fig. 1c. Copyright (2011) by the Nature)

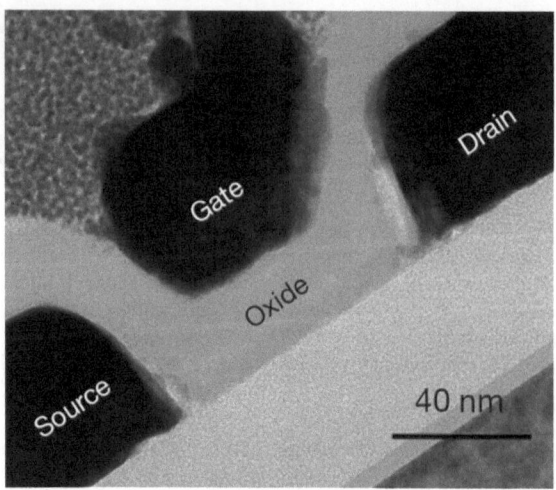

have demonstrated FETs of better characteristics by going back to the micro-mechanically cleaved graphene, not a practical method. A different method for obtaining higher quality graphene on a large substrate has been explored by Wu et al. (2011). This is based on CVD growth of graphene on Cu, with subsequent transfer of large areas (cm scale) of graphene onto diamond-like-carbon (DLC) substrates. These two innovations led to cutoff frequencies as high as 155 GHz at room temperature with gate length of 40 nm. The device of Wu et al. is imaged in Fig. 5.4, with summary of several devices in Fig. 5.5.

The procedure of Wu et al. (2011) follows the method of Li et al. (2009), for the graphene growth.[1]

The CVD graphene, after transfer to to diamond-like-carbon (DLC) coated surface, was characterized by Raman spectroscopy before device fabrication. Diamond-like-carbon DLC film was grown on an 8-in. Si substrate using cyclo-hexane (C_6H_{12}) with a vapor pressure of 1.8 psi in a CVD chamber. The flow rate was typically 25–40 cc STP per min at 100 milliTorr pressure. The DLC growth rate is 32 A/s at 60 °C, and this was followed by anneal at 400 °C for 4 h. The source and drain electrodes were 20 nm of Pd followed by 30 nm Au using e-beam evaporation. The gate was an oxidized Al layer, evaporated by electron beam deposition, followed by 15 nm atomic layer deposition (ALD) of Al_2O_3. The authors state that the DLC support for the CVD graphene is superior to the tra-ditional oxidized Si support. They attribute the superiority to the higher phonon

---

[1] Namely, after evacuation of the chemical vapor deposition CVD chamber, a Cu foil was heated to 875 °C in forming gas (H_2/Ar) for 30 min. After this reduction, the Cu foil was exposed to ethylene at 975 °C for 10 min and then cooled. PMMA was spin-coated on top of the graphene layer that had formed on one side of the Cu foil. The Cu foil was then dissolved in 1 M iron chloride solution. The remaining graphene/PMMA layer was washed and transferred to the desired substrate. Subsequently the PMMA was dissolved by treatment with hot acetone for 1 h.

Fig. 5.5 Gate-length-dependence of cut-off frequency. Summary of cut-off frequencies for family of graphene FET devices with varying gate lengths. (Reprinted figure with permission from Wu et al., Fig. 3b. Copyright (2011) by the Nature)

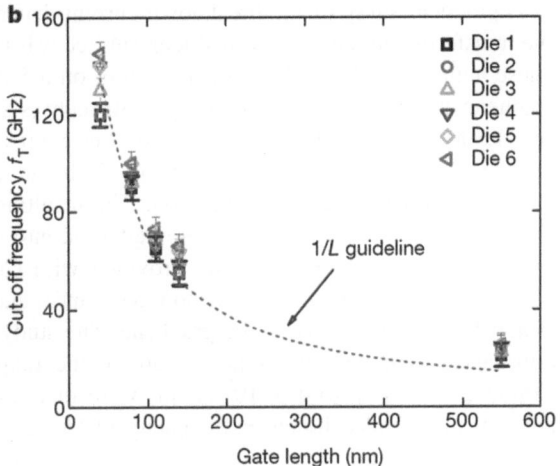

energy in diamond (165 meV) and a lower surface trap density than SiO_2, since DLC is non-polar and chemically inert.

High-frequency voltage amplifiers were fabricated by Han et al. (2011) using CVD-grown graphene transferred to a Si wafer. The devices, with 500 nm gate lengh, showed 5 dB low-frequency gain and 3 dB bandwidth greater than 6 GHz.

Wafer-scale graphene integrated circuits have recently been reported by Lin et al. (2011b), again based on SiC growth of graphene. These circuits include graphene FETs and inductors, and perform as broadband radio-frequency mixers at frequencies up to 10 GHz.

5.2 Interconnects on Chips

Metal wires or interconnects are an essential aspect of silicon chip design and manufacture, and graphene is a conductor much superior to any metal. Silicon chips commonly have six layers of metallization. Much publicity was attached to the change from Al to Cu interconnects in chip manufacture several years ago (Rosenberg et al. 2000). A comparative analysis of graphene interconnects was offered by Xu, et al. (2009). The more recent report of Bae et al. (2010) of the low sheet resistance of 30 Ω/square, measured on a four-layer p-doped graphene sheet, may be encouraging from the point of view of making on-chip interconnects. The growth of the semiconductor industry is paced by ITRS, the International Technology Roadmap for Semiconductors, that predicts that by 2020 the basic width of wiring will be 22 nm, and will require current density for interconnects of 5.8×10^6 A/cm². This current density is regarded as impossible in Cu wiring, making clear a definite opportunity for graphene interconnects.

A practical study of the breakdown-current-density in multilayer CVD graphene transferred onto an oxidized silicon surface was reported by Lee et al. (2011a). These authors used a 500 nm thick-Ni film on a Si/SiO$_2$ substrate to grow CVD graphene at 1,000 °C in 10 min using 5–30 sccm (standard cc per min) of methane and 1,300 sccm of hydrogen at atmospheric pressure. It was found that the thicknesses of the graphene layers, measured by atomic force microscopy (AFM), increased with methane concentration. The resulting films were transferred to a clean oxidized Si substrate, and were patterned into wires of 1 and 10 μm widths, and lengths from 2 to 1,000 μm, provided with Cr/Au or Ti/Au contacts. The contacts were determined to have low resistance, and measurements on the wires established the resistivity of the graphene. The study emphasized 10–20 nm thick films and found resistances per square in the range 500–1,000 Ω. Breakdown current densities up to 4×10^7 A/cm^2 were measured, exceeding, by at least an order of magnitude, the current capacity of Cu. The breakdown mechanism was identified as heating.

Growth directly on silicon would be desirable from a production point of view, but using present methods would limit the performance of the graphene. (As mentioned earlier, Hackley et al. (2009) have had little success growing graphitic carbon directly on Si 111.) Higher values of conductivity have recently been reported by Jain et al. (2012) for graphene on hexagonal BN substrates. The graphene is grown by CVD on Cu foil and then transferred to hexagonal BN mechanically-exfoliated flakes that, in turn, had been transferred onto an oxidized Si wafer. The electrical characteristics of Cu-grown CVD graphene on this surface are excellent, including $\mu = 1.5$ m^2/Vs at carrier density 1×10^{12} cm^{-2}. The breakdown current density is recorded as 1.4×10^9 A/cm^2 as shown in Fig. 5.6, in comparison to Cu-CVD grown graphene placed on SiO$_2$, and to mechanically-exfoliated graphene on SiO$_2$.

In chip manufacture, interconnects are central and ubiquitous, there are typically five layers of such wiring. It would seem that this is an application where the Cu substrate layer, underlying CVD graphene, need not be removed, and that a wafer-size graphene polycrystal might be transferred onto the Si wafer, patterned and etched, to leave one layer of interconnect wiring. In CVD growth it is likely that multiple layers of graphene on the Cu film could be achieved to reduce the resistance per square, without appreciably raising the cost or complexity of the process. The catalytic Cu film, in principle, can be an evaporated film, rather than a foil, to appreciably reduce its thickness and cost.

An extensive report demonstrating the use of graphene interconnects for arrays of inorganic microscale light-emitting diodes on stretchable rubber substrates has been given by Kim et al. (2011a). These authors find that linear distortions up to 100 % can be accommodated in a properly designed system.

Fig. 5.6 Current density versus voltage for monolayer graphene on (*top* to *bottom*) h-BN, CVD graphene on SiO$_2$, and mechanically exfoliated graphene on SiO$_2$. Jain et al. (2012), Fig. 3a. (Reprinted figure with permission from Jain et al., Fig. 3a. Copyright (2012) by the IEEE)

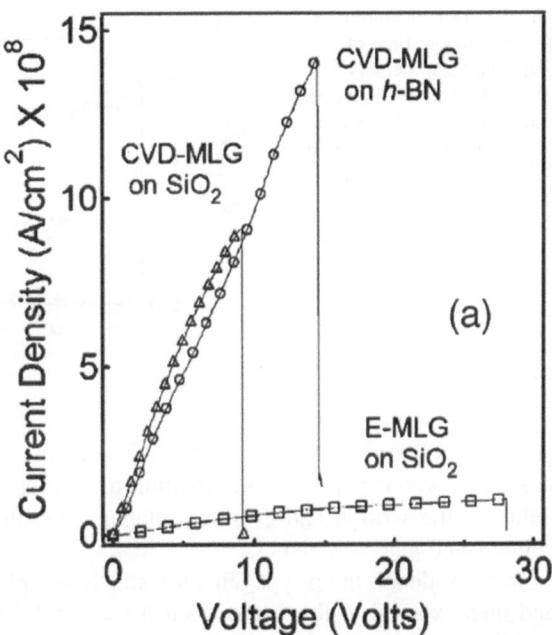

5.3 Flash Memory Cells

A flash memory cell is, essentially, a storage capacitor of minimum lateral dimension, currently in the vicinity of 45 nm, with read- and write-capabilities, and storage time for charge measured in years. High performance laptop computers use flash memory in place of magnetic disk memory, and USB flash-memory drives are ubiquitous. The current state-of-the-art in flash memory is the polysilicon floating-gate device on a p-type Si wafer. (A review is offered by Lu et al. 2009).

A promising graphene flash memory device has been described by a consortium of workers at IBM, UCLA, Samsung and Aerospace Corp (Hong et al. 2011). (This paper also gives reference to earlier approaches to non-volatile memory based on graphene.) These authors have patterned capacitor-like storage devices on large-area CVD-grown graphene transferred onto an oxidized p-type Si substrate. Individual capacitor devices of form (graphene/5 nm SiO$_2$ tunnel oxide/p-type Si) have been shown by the authors to have a charge retention time of 10 years for 8 % loss of charge (see Fig. 5.7). The height of the graphene/quartz barrier is about 3 eV. The storage capacitor is charged from an upper Gate electrode, separated by a gate barrier. The upper electrode (graphene) of the storage capacitor is covered by 35 nm of sapphire (Al$_2$O$_3$), with a write/read gate-electrode on the top. The devices are reported to have a window of voltage on the read/write electrode of ±6 V for secure charge storage, with program/erase voltages at ±7 V. The authors have determined that the stored charge resides on the graphene layer. The fabrication of the devices requires extreme care with respect to the properties

Fig. 5.7 Demonstration of
10-year storage time of
graphene flash memory
device. (Reprinted figure with
permission from Hong et al.,
Fig. 4a. Copyright (2011) by
the American Chemical
Society)

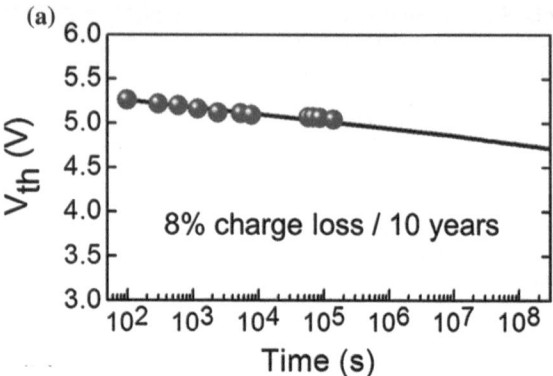

of the tunnel barrier of SiO_2 and the gate-electrode barrier of Al_2O_3. The care is
needed to preserve the desired retention time of charge and to achieve reproduc-
ibility in the writing and erasing voltages, controlled, respectively, by the tunneling—and gate—barriers.

In more detail, the p-type silicon wafer is completely cleaned of native oxide,
and then exposed to flowing oxygen for 7 s at 1,000° C. The graphene layer or
multilayer, grown on Cu foil (or on deposited Ni), was then transferred onto the
carefully oxidized surface. The gate oxide on the graphene was prepared by initial
deposit of 1.1 nm of Al, oxidized in air for 2 days, followed by 300 cycles (layers)
of Al_2O_3 by atomic layer deposition (ALD). The gate electrode on this oxide
barrier is Ti/Al/Au (10/500/10 nm) using photolithography and electron-beam
evaporation. The gate electrodes were varied in area $(2.5 \times 10^{-5}$–
7.4×10^{-4} cm$^2)$. These electrodes were used as a mask to etch the individual gate
stacks. The Al_2O_3 and graphene, outside the device area, are removed with 30 s.
Cl_2 reactive–ion-etching, followed by 3 min of O_2 plasma.

The packing density of flash memory cells is limited by interference voltage
between neighboring cells, termed crosstalk. The crosstalk interference voltage is
defined as shift in threshold voltage of an unprogrammed cell by its two nearest-
neighbor cells. The essential feature leading to smaller crosstalk is the vertical
height of the cell, that controls the inter-cell capacitance. In principle, an evaporated metal can be made thin, but in practice it may break into islands and not
conduct well. The covalent bonding of the graphene is needed to achieve a stable,
diffusion-free, and electrically continuous conductive thin layer, as small as
0.34 nm in thickness, but here assumed to be 1 nm (graphene trilayer) in Fig. 5.8.

The authors conclude that the graphene-based flash memory (GFM) has significant advantages in miniaturization below the 25 nm "node" in semiconductor
scaling, based on the lower interference, and also in lower operating voltage to
achieve the minimum window of stable storage voltage, 1.5 V accepted in current
practice. The energy per bit is predicted to be reduced by 75 % in the GFM
graphene flash memory device. The advantages of graphene in this application are
the high work function, the high density of states (relative to polysilicon), and,

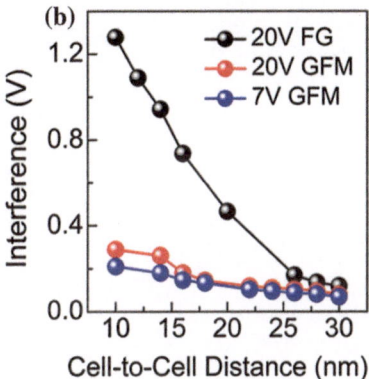

Fig. 5.8 Superiority of graphene flash memory limits of flash memory (FM) cell packing from cross-talk, comparing simulations of traditional FG floating-gate polysilicon flash memory (*upper curve*) with Graphene Flash Memory (GFM) of tri-layer graphene in *lower two curves*. Interference voltage is defined as shift in threshold voltage of unprogrammed cell by its two nearest–neighbor cells. The superiority of GFM in this regard, below cell spacing of about 25 nm, comes from smaller inter-cell capacitance consequent to the lower cell height. This is an advantage in miniaturization of graphene flash memory versus traditional polysilicon flash memory. (Reprinted figure with permission from Hong et al., Fig. 5b. Copyright (2011) by the American Chemical Society)

primarily, the vanishing vertical height, a few multiples of 0.34 nm, for a continuous conductive layer. Thus it appears that graphene flash memory is promising as a substitution for existing silicon memory.

In summary, in this Chapter we have seen several applications for graphene within the existing semiconductor technology. Continuing improvement in fabrication methods for graphene will likely allow several of these to be implemented.

References

Affoune A, Prasad B, Sato H, Enoki T, Kaburagi Y, Hishiyama Y (2001) Chem Phys Lett 348:17
Appenzeller J, Lin Y-M, Knoch J, Avouris P (2004) Phys Rev Lett 93:196805
Bae S, Kim Y, Lee Y, Xu X, Park J-S, Zheng Y, Balakrishnan J, Lei T, Kim H, Song Y,
    Kim Y-LJ, Kim K, Ozyllmaz B, Ahn J-H, Hong B, Iijima S (2010) Nat Nanotechnol 5:574
Bai J, Zong X, Jiang S, Huang Y, Duan X (2010) Nat Nanotechnol 5:190
Balandin A, Ghosh S, Bao W, Calizo I, Teweldebrhan D, Miao T, Lau C (2008) Nano Lett 8:902
Balestra F (2010) Silicon-based devices and materials for nanoscale CMOS and beyond-CMOS.
    In: Luryi S, Xu J, Zaslavsky J (eds) Future trends in microelectronics: from nanophotonics to
    sensors and energy. IEEE Press, Wiley, Hoboken, pp 109–126 (Section 2.1)
Berger C, Song Z, Li T, Li X, Oghazghi A, Feng R, Dai Z, Marchenkov A, Conrad E, First P,
    de Heer W (2004) J Phys Chem B 108:19912
Berger C, Song Z, Li S, Wu X, Brown N, Naud C, Mayou D, Hass J, Marchenkov A, Conrad E,
    First P, de Heer W (2006) Science 312:1191
Bolotin K, Sikes K, Hone J, Stormer H, Kim P (2008) Phys Rev Lett 101:096802
Bonnani A, Bobisch C, Moller R (2008) Rev Sci Instrum 79:83704
Bosshard P (2006) Global climate and energy project (GCEP): an assessment of solar energy
    conversion technologies and research opportunities. In: Global climate and energy project,
    Stanford University, Stanford
Bostwick A, McChesney J, Ohta T, Rotenberg E, Seyller T, Horn K (2009) Prog Surf Sci 84:380
Britnell L, Gorbachev R, Jalil R, Belle B, Schedin F, Mishcenko A, Georgiou T, Katsnelson M,
    Eaves L, Morozov S, Peres N, Leist J, Geim A, Novoselov K, Ponomarenko L (2012) Science
    335:947
Brodie B (1859) Phil Trans Roy Soc Lond 149:249
Brown SF (2013) Carbon fiber, light and strong, arrives where it's most needed. New York
    Times, New York
Bunch J, van der Zande A, Verbridge S, Frank I, Tanenbaum D, Parpia J, Craighead H, McEuen P
    (2007) Science 315:490
Bunch J, Verbridge S, Alden J, van der Zande A, Parpia J, Craighead H, McEuen P (2008) Nano
    Lett 8:2458
Bunyk P, Likharev K, Zinoviev D (2001) Int J High Speed Electron Syst 11:257
Cao H, Yu Q, Jauregui L, Tian J, Wu W, Liu Z, Jalilian R, Benjamin D, Jiang Z, Bao J, Pei S,
    Chen Y (2010) Appl Phys Lett 96:122106
Castro Neto A, Guinea F, Peres N, Novoselov K, Geim A (2009) Rev Mod Phys 81:109
Cheianov V, Fal'ko V (2006) Phys Rev B 74:041403
Chen W, Chen H, Lan H, Schulze T, Zhu W, Zhang Z (2012) Phys Rev Lett 109:265507
Choucair M, Thordarson P, Stride J (2008) Nat Nanotechnol 4:30
Chuang A, Robertson J, Boskovic B, Koziol K (2007) Appl Phys Lett 90:123107
DaSilva A, Zou K, Jain J, Zhu J (2010) Phys Rev Lett 104:236601

Dai H, Wong E, Lieber C (1996) Science 272:523

Dato A, Radmilovic V, Lee Z, Phillips J, Frenklach M (2008) Nano Lett 8:2012

Dean C, Young A, Meric L, Lee C, Wang L, Sorgenfrei S, Watanabe K, Taniguchi T, Kim P, Shepard K, Hone J (2010) Nat Nanotechnol 5:722

Delucchi M, Jacobsen M (2011) Energy Policy 39:1170

Deng D, Pan X, Yu L, Cui Y, Jiang Y, Qi J, Li W-X, Fu Q, Ma X, Xue Q, Sun G, Bao X (2011) Chem Mater 23:1188

Dikin D, Stankovich S, Zimney E, Piner R, Dommett G, Evmenenko G, Nguyen S, Ruoff R (2007) Nature 448:457

DiVincenzo D, Mele E (1984) Phys Rev B 29:1685

Dresselhaus M, Dresselhaus G (1981) Adv Phys 30:139

Du X, Skachko I, Barker A, Andrei E (2008) Nat Nanotechnol 3:491

Eda G, Lin Y-Y, Miller S, Chen C-W, Su W-F, Chhowalla M (2008) Appl Phys Lett 92:233305

Efetov D, Kim P (2010) Phys Rev Lett 105:256805

Ekinci K, Roukes M (2005) Rev Sci Instrum 76:61101

Elias D, Gorbachev R, Mayorov A, Morozov S, Zhukov A, Blake P, Ponomarenko L, Grigorieva I, Novoselov K, Guinea F, Geim A (2011) Nat Phys 7:701

Emtsev K, Bostwick A, Horn K, Jobst J, Kellogg G, Ley L, McChesney J, Ohta T, Reshanov S, Rohrl J, Rotenbert E, Schmid A, Waldmann D, Weber H, Seyller T (2009) Nat Mater 8:203

Esaki L (1958) Phys Rev 109:503

Ferain I, Colinge C, Colinge J-P (2011) Nature 479:310

Ferrari A, Meyer J, Scardaci V, Casiraghi C, Lazzeri M, Mauri F, Pascanee S, Jian D, Novoselov K, Roth S, Geim A (2006) Phys Rev Lett 97:187401

Fiori G, Iannaccone G (2009) IEEE Electron Device Lett 30:261

Fiori G, Iannaccone G (2009) IEEE Electron Device Lett 30:1096

Fiori G, Betti A, Bruzzone S, D'Amico P, Iannaccone G (2011) IEDM Tech Digest, 11.4.1–11.4.4, ISSN: 0163-1918 (Washington, DC, USA)

Fiori G, Betti A, Bruzzone S, Iannaccone G (2012) ACS Nano 6:2642

Forbeaux I, Themlin J, Langlais V, Yu M, Belkhir H, Debever J (1998) Surf Rev Lett 5:193

Frank D, Taur F, Wong H-S (1998) IEEE Electron Device Lett 19:385

Franklin R (1951) Proc Roy Soc A 209:196

Franklin A, Luisier M, Han S-J, Tulevski G, Breslin C, Gignac L, Lundstrom M, Haensch W (2012) Nano Lett 12:758

Gall N, Mikhailov S, Ruf'kov E, Tontegode A (1987) Surf Sci 191:185

Gall N, Ruf'kov E, Tontegode A (1997) Int J Mod Phys B 11:1865

Gamby J, Taberna P, Simon P, Fauvarque J, Chesneaut M (2001) J Power Sources 101:109

Geim A (2011) Angew Chem Int Ed 50:6966

Geim A (2011) Rev Mod Phys 83:851

Gomez-Navarro C, Weitz R, Bittner A, Scolari M, Mews A, Burghard M, Kern K (2007) Nano Lett 7:3499

Graetzel M (2001) Nature 414:138

Grant J, Haas T (1970) Surf Sci 21:76

http://europa.eu/rapid/press-release_IP-13-54_en.htm (1$B funding graphene research Prof. Jan Kinaret, Chalmers University, Sweden, Jan 2013)

http://www.nanowerk.com/spotlight/spotid=25744.php

Hackley J, Ali D, DiPasquaale J, Demaree J, Richardson C (2009) Appl Phys Lett 95:133114

Han M, Ozyilmaz B, Zhang Y, Kim P (2007) Phys Rev Lett 98:206805

Han S-J, Jenkins K, Garcie A, Frandkon A, Bol A, Haensch W (2011) Nano Lett 11:3690

Harigaya K, Enoki T (2002) Chem Phys Lett 351:128

Harris P, Liu Z, Suenaga K (2008) J Phys Condens Matter 20:362201

Hass J, Barchon F, Millan-Otoya J, Sprinkle M, Sharma H, de Heer W, Berger C, First P, Magaud L, Conrad E (2008) Phys Rev Lett 100:125504

Hernandez Y, Nicolosi V, Lotya M, Blighe R, Sun Z, De S, McGovern I,Holland B, Byrne M, Gun'ko Y, Boland J, Niraj P, Duesberg G, Krishnamurthy S, Goodhue R, Hutchison J, Scardacci V, Ferrari A, Coleman J (2008) Nat Nanotechnol 3:563

Hicks J, Tejeda A, Taleb-Ibrahim A, Nevius M, Wang F, Shepperd, K, Palmer J, Bertran F, Le Fevre P, Kunc J, de Heer W, Berger C, Conrad E (2012) Nat Phys. doi:10.1038/NPHYS52487

Hiraoka T, Izadi-Najafabadi A, Yamada T, Futaba D, Yasuda S, Tanaike O, Hatori H, Yumura M, Iijima S, Hata K (2010) Adv Funct Mater 20:422

Hong A, Song E, Yu H, Allen M, Kim J, Fowler J, Wassel J, Park Y, Wang ZJ, Kaner R, Weiller B, Wang K, (2011) ACS Nano 5:7812

Homoth J, Wenderoth M, Druga T, Winking L, Ulbrich R, Bobisch C, Weyers B, Bonnani A, Zubkov E, Bernhart A, Kaspers M, Moller R (2009) Nano Lett 9:1588

Hummers W, Offerman R (1958) J Am Chem Soc 80:1339

Ionescu A, Riel H (2011) Nature 479:329

Jain N, Bansal T, Durcan C, Yu B (2012) IEEE Electron Device Lett 33:925

Jeon I-Y, Shin Y-R, Sohn G-J, Choi H-U, Bae S-Y, Mahmood J, Jung S-M, Seo J-M, Kim M-J, Chang D, Dai L, Baek J-B (2012) Proc Natl Acad Sci 109:5588

Ji S-H, Hannon J, Tromp R, Perebeinos V, Tersoff J, Ross F (2012) Nat Mater 11:114

Jia K, Hofmann M, Meunier V, Sumpter B, Campos-Delgado J, Romo-Herrera J, Son H, Hsieh Y, Reina A, Kong J, Terrones M, Dresselhaus M (2009) Science 323:1701

Jo G, Choe M, Cho C-Y, Kim J, Park W, Lee S, Hong W-K, Kim T-W, Park S-J, Hong B, Kahng Y, Lee T (2010) Nanotechnology 21:175201

Jo G, Choe M, Lee S, Park W, Kahng Y-H, Lee T (2012) Nanotechnology 23:112001

Kane E (1959) J Phys Chem Solids 12:181

Karim S, Maaz K, Ensinger W (2009) J Phys D Appl Phys 42:185403

Katsnelson M, Novoselov K, Geim A (2006) Nat Phys 2:620

Kim K, Zhao Y, Jang H, Lee S, Kim J, Kim K, Ahn J-H, Kim P, Choi J-Y, Hong B (2009) Nature 457:706

Kim P, Lee Z, Regan W, Kisielowski C, Crommie M, Zettle A (2011) ACS Nano 5:2142

Kim R-H, Bae M-H, Kim D, Cheng H, Kim B, Kim D-H, Li M, Wu J, Du F, Kim H-S, Kim S, Estrada D, Hong S, Huang Y, Pop E, Rogers J (2011) Nano Lett 11:3881

Klein O (1929) Z Phys 53:157

Knoch J, Mantl S, Appenzeller J (2007) Solid State Electron 51:572

Kobayashi K, Tanimura M, Makai H, Yoshimura A, Yoshimura H, Kojima K, Tachibana M (2007) J Appl Phys 101:094306

Koch M, Ample F, Joachim C, Grill L (2012) Nat Nanotechnol 7:713

Koga J, Toriumi A (1997) Appl Phys Lett 70:2138

Krumhansl J, Brooks H (1953) J Chem Phys 21:1663

Landau L, Lifshitz E (1986) Theory of elasticity, 3rd edn. Pergamon Press, Oxford

Lee J-Y, Connor S, Cui S, Peumans P (2008) Nano Lett 8:692

Lee W, Park J, Sim S, Jo S, Kim K, Hong B, Cho K (2011) Adv Mater 23:1752

Lee K-J, Chandrakasan A, Kong J (2011) IEEE Electron Device Lett 32:557

Levendorf M, Kim C-J, Brown L, Huang P, Havener R, Muller D, Park H (2012) Nature 488:627

Li X, Wang X, Zhang L, Lee S, Dai H (2008) Science 319:1229

Li X, Cai W, An J, Kim S, Nah J, Yang D, Piner R, Velamakanni A, Jung I, Tutuc E, Banerjee S, Colombo L, Ruoff R (2009) Science 324:1312

Li X, Magnuson C, Venugopal A, Tromp R, Hannon J, Vogel E, Colombo L, Ruoff R (2011) J Am Chem Soc 133:2816

Li Z, Wu P, Wang C, Fan X, Zhang W, Zhai X, Zeng C, Li Z, Yang J, Hou J (2011) ACS Nano 5:3385

Liang X, Jung Y-S, Wu S, Ismach A, Loynick D, Cabrini S, Bokor J (2010) Nano Lett 10:2454

Liao L, Lin Y-C, Bao M, Cheng R, Bai J, Liu Y, Qu Y, Wang K, Huang Y, Duan X (2010) Nature 467:305

Lin Y, Dimitrakopoulos C, Jenkins K, Farmer D, Chiu H-Y, Grill A, Avouris Ph (2010) Science 327:662.

Lin Y, Valdes-Garcia A, Han S-J, Farmer D, Meric I, Sun Y, Wu Y, Dimitrakopoulos C, Grill A, Avouris P, Jenkins K (2011) Science 332:1294

Lin T, Huang F, Liang J, Wang Y (2011) Energy Environ Sci 4:862

Liu C, Yu Z, Neff D, Zhamu A, Jang Z (2010) Nano Lett 10:4863

Low T, Perebeinos V, Tersoff J, Avouris P (2012) Phys Rev Lett 108:096601

Lu C-Y, Hsieh K-Y, Liu R (2009) Microelectron Eng 86:283

Malard L, Pimnta M, Dresselhaus G, Dresselhaus M (2008) Phys Rep 473:51

Mao Y, Wang W, Wei D, Kaxiras E, Sodroski J (2011) ACS Nano 5:1395

Marconcini P, Cresti A, Triozon F, Fiori G, Biel B, Niquet Y-M, Macucci M, Roche S (2012) ACS Nano 6:7942

Mayorov A, Gorbachev R, Morozov S, Britnell L, Jalil R, Ponomarenko L, Blake P, Novoselov K, Watanabe K, Taniguchi T, Geim A (2011) Nano Lett 11:2396

McCann E (2006) Phys Rev B 74:161403

McEuen PL, Bockrath M, Cobden DH, Yoon Y-L, Louie SL (1999) Phys Rev Lett 83:5098

McKenzie D, Muller D, Pailthorpe B, Wang Z, Kravtchinskaia D, Segal D, Lukins P, Martin P, Amaratunga G, Gaskell P, Saeed A (1991) Diam Relat Mater 1:51

Miao X, Tongay S, Petterson M, Berke K, Rinzler A, Appleton B, Hebard A (2012) Nano Lett 12:2745

Mkhoyan K, Contryman A, Silcox J, Stewart D, Eda G, Mattevi C, Miller S, Chhowalla M (2009) Nano Lett 9:1058

Mounet N, Marzari H (2005) Phys Rev B71:205214

Muller S, Holzer F, Arai H, Haas O (1999) J New Mater Electrochem Syst 2:227

Nagashima A, Nuka A, Itoh H, Ichinokawa T, Oshima C, Otani S (1993) Surf Sci 291:93

Nair R, Blake P, Grigorenko A, Novoselov K, Booth T, Stauber T, Peres N, Geim A (2008) Science 320:1308

Nicklow R, Wakabayashi N, Smith G (1972) Phys Rev B5:4751

Novoselov KS, Geim AK, Morozov SV, Jiang D, Zhang Y, Dubonos SV, Grigorieva IV, Firsov AA (2004) Science 306:666

Novoselov K, Jiang Z, Zhang Y, Morozov S, Stormer H, Zeitler U, Maan J, Boebinger G, Kim P, Geim A (2007) Science 315:1379

Novoselov K (2011) Rev Mod Phys 83:837

Novoselov K, Fal'ko V, Colombo L, Gellert P, Schwab M, Kim K (2012) Nature 490:192

Ohta T, Bostwick A, Seyller T, Horn K, Rotenberg E (2006) Science 313:951

Park S, Ruoff R (2009) Nat Nanotechnol 4:217

Ponomarenko L, Yang R, Gorbachev R, Blake P, Mayorov S, Novoselov K, Katsnelson MM, Geim A (2010) Phys Rev Lett 105:136801

Ponomarenko L, Geim A, Zhukov A, Jalil R, Morozov S, Novoselov K, Grigorieva I, Hill E, Cheianov V, Fal'ko F, Watanabe K, Taniguchi T, Gorbachev R (2011) Nat Phys 7:958

Pop E, Mann D, Wang Q, Goodson K, Dai H (2006) Nano Lett 6:96

Pureval J (2010) Potassium intercalated graphene, Chapter 2 in thesis. http://thesis.library.caltech.edu/5574/4/Purewal_Thesis_Ch2.pdf.

Quinn J, Kawamoto G, McCombe B (1978) Surf Sci 73:190

Ramon M, Gupta A, Corbet C, Ferrer D, Movva C, Carpenter G, Colombo L, Bourianoff G, Docy M, Akinwande D, Tutuc E, Banerjee S (2011) ACS Nano 5:7198

Reddick W, Amaratunga A (1995) Appl Phys Lett 67:494

Reina A, Son H, Jiao L, Fan B, Dresselhaus M, Liu Z, Kong J (2008) J Phys Chem C 112:17741

Reina A, Jia X, Ho J, Nezich D, Son H, Bulovic V, Dresselhaus M, Kong J (2009) Nano Lett 9:30

Rosenberg K., Edelstein D, Hu C-K, Rodbell K (2000) Annu Rev Mater Sci 30:229

Rosenstein B, Lewkowicz M, Maniv T (2013) Phys Rev Lett 110:066602

Ruess G, Vogt F (1948) Monatsh Chem 78:222

Ruoff R (2008) Nat Nanotechnol 3:10

Sivudu SK, Mahajan Y (2012) Nanotechnol Insights 3:6

Schedin F, Geim A, Morozov S, Hill E, Blake F, Katsnelson M, Novoselov K (2007) Nat Mater 6:652

Schiffer P, O'Keefe M, Osheroff D, Fukuyama H (1993) Phys Rev Lett 71:140

Schniepp H, Li J-L, McAllister M, Sai H, Herrera-Alonso M, Adamson D, Prud'homme R, Car R, Saville D, Aksay I (2006) J Phys Chem B 110:8535

Schwierz F (2010) Nat Nanotechnol 5:487

Seabaugh A, Zhang Q (2010) Proc IEEE 98:2095

Segal M (2009) Nat Nanotechnol 4:612

Semenoff G (1984) Phys Rev Lett 53:2449

Seol J, Jo I, Moore A, Lindsay L, Aitken Z, Pettes M, Li X, Yao Z, Huang R, Broido D, Mingo N, Ruoff R, Shi L (2010) Science 328:213

Seyller T (2012) Epitaxial graphene on SiC(0001). In: Raza H (ed) Graphene nanoelectronics: metrology, synthesis, properties, and applications. Springer Verlag, Berlin, pp 135–15099

Shi E, Li H, Yang L, Zhang L, Li Z, Li P, Shang Y, Wu S, Li X, Wei J, Wang K, Zhu H, Wu D, Fang Y, Cao A (2013) Nano Lett 13:1776

Shioyama H (2001) J Mate Sci Lett 20:499

Shivaraman S, Barton R, Yu X, Alden J, Herman L, Chandrashekhar M, Park J, McEuen P, Parpia J, Craighead H, Spencer J (2009) Nano Lett 9:3100

Shulaker M, Hills G, Patil N, Wei H, Chen H-Y, Wong H-S, Mitra S (2013) Nature 501:526

Simmons J (1963) J Appl Phys 34:1793

Singh D, Iyer P, Giri P (2011) Int J Nanosci 10:1

Sojoudi H, Baltazar J, Tolbert L, Henderson C, Graham S (2012) ACS Appl Mater Interfaces 4:4781

Sprinkle M, Siegel D, Hu Y, Hicks J, Tejeda A, Taleb-Ibrahimi A, Le Fevre P, Bertran F, Vizzini S, Enriquez H, Chiang S, Soukiassian P, Berger C, de Heer W, Lanzara A, Conrad E (2009) Phys Rev Lett 103:226803

Sprinkle M, Ruan M, Hu Y, Hankinson J, Rubio-Roy M, Zhang B, Wu X, Berger C, de Heer W (2010) Nat Nanotechnol 5:727

Stankovich S, Dikin D, Dommett G, Kohlhaas K, Zimney E, Stach E, Piner R, Nguyen S, Ruoff R (2006) Nature 442:282

Stankovich S, Dikin D, Piner R, Kohlhaas K, Kleinhammes A, Jia Y, Wu Y, Nguyen S, Ruoff R (2007) Carbon 45:1558

Staudenmaier L (1898) Ber Dtsch Chem Ges 31:1481

Stoller M, Park S, Zhu Y, An J, Ruoff R (2008) Nano Lett 8:3498

Subrahmanyam K, Kumar P, Maitra U, Govidaraj A, Hembram K, Waghmare U, Rao C (2011) Proc Natl Acad Sci 108:2674

Szafranek B, Fiori G, Schall D, Neumaier D, Kurz H (2012) Nano Lett 12:1324

Tani S, Blanchard F, Tanake K (2012) Phys Rev Lett 109:166603

Tao L, Lee J, Chou H, Holt M, Ruoff R, Akinwande D (2012) ACS Nano 6:2319

Tsen A, Brown L, Levendorf M, Ghahari F, Huang P, Havener R, Ruiz-Vargas C, Mutter D, Kim P, Park J (2012) Science 336:1143

Tung VC, Allen MJ, Yang Y, Kaner RB (2009) Nat Nanotechnol 4:25

Van Bommel A, Crombeen J, Van Tooren A (1975) Surf Sci 48:463

Vogel FL (1977) J Mater Sci 12:982

Wakabayashi H, Ezaki T, Sakamoto T, Kawaura H, Ikarishi N, Ikezawa N, Narihiro M, Ochiai Y, Ikezawa T, Takeuchi K, Yamamoto T, Hane M, Mogami T (2006) IEEE Trans Electron Devices 53:1961

Wallace P (1947) Phys Rev 71:622

Wang J, Zhu M, Outlaw R, Zhao X, Manos D, Holloway B (2004) Appl Phys Lett 85:1265

Wang P, Hilsenbeck K, Nirschl T, Oswald M, Stepper C, Weis M, Schmitt-Landsiedel D, Hansch W (2004) Solid State Electron 48:2281

Wang W, Poudel B, Wang D, Ren Z (2005) J Am Chem Soc 127:18018

Wang X, Ouyang Y, Li X, Wang H, Guo J, Dai H (2008) Phys Rev Lett 100:206803

Wang X, Zhi L, Mullen K (2008) Nano Lett 8:323

Wolf E (1985) Principles of electron tunneling spectroscopy. Oxford University Press, Oxford (See also 2nd Edition, 2012)

Wolf E (2014) Graphene: a new paradigm in condensed matter and device physics. Oxford University Press, Oxford

Woody T (2009) U. S. Company and China Plan Solar Project, New York Times, New York. (See also Yergin D (2011) The Quest. Penguin, New York, pp 581)

Wu J, Becerril H, Bao Z, Liu Z, Chen Y, Peumanns P (2008) Appl Phys Lett 92:263302

Wu J, Agrawal M, Becerril H, Bao Z, Liu Z, Chen Y, Peumans P (2010) ACS Nano 4:43

Wu Y, Lin Y-M, Bol A, Jenkins K, Xia F, Farmer D, Zhu Y, Avouris Ph (2011) Nature 472:74

Xu C, Li H, Banerjee K (2009) IEEE Trans Electron Devices 56:1567

Xue J, Sanchez-Yamagishi J, Bulmash D, Jacquod P, Deshpande A, Watanabe K, Taniguchi T, Jarillo-Herrero P, LeRoy B (2011) Nat Mater 10:282

Yang S, Feng X, Ivanovici K, Mullen K (2010) Angew Chem Int Ed 49:8408

Yao Z, Kane C, Dekker C (2000) Phys Rev Lett 84:2041

Yoo E, Kim J, Hosono E, Zhou H-S, Kudo T, Honma I (2008) Nano Lett 8:2277

Young A, Kim P (2009) Nat Phys 5:222

Zacharia R, Ulbricht H, Hertel T (2004) Phys Rev B 69:155406

Zakharchenko K, Fasolino A, Los J, Katsnelson M (2011) J Phys Condens Matter 23:202202

Zener C (1934) Proc Roy Soc A145:523

Zhang Y, Tan Y-T, Stormer HL, Kim P (2005) Nature 438:201

Zhao X, Outlaw R, Wang J, Zhu M, Smith D, Holloway B (2006) J Chem Phys 124:194704

Zheng Y, Ni G-X, Toh C-T, Tan C-Y, Yao K, Ozyilmaz B (2010) Phys Rev Lett 105:166602

Zhu Y, Murali S, Stoller M, Ganesh J, Cai W, Ferreira P, Pirkle A, Wallace R, Cychosz K, Thommes M, Su D, Stach E, Ruoff R (2011) Science 332:1537